Adventures in Earth and Environmental Science

Practical Manual
Book 1

Dr. Peter T. Scott

First released 2018 all rights reserved Felix Publishing
Felix Publishing 2018

Felix Publishing 2018
www.felixpublishing.com.au
email:
info@felixpublishing.com

Print copies available from publisher.

ADVENTURES in EARTH and ENVIRONMENTAL SCIENCE PRACTICAL MANUAL Book 1

Also by the author:

ADVENTURES in EARTH and ENVIRONMENTAL SCIENCE Book 1

ADVENTURES in EARTH and ENVIRONMENTAL SCIENCE Book 2

    and companion PRACTICAL MANUAL for Book 2.

ADVENTURES IN EARTH and ENVIRONMENTAL SCIENCE

    (the composite book containing books 1 and 2)

    and accompanying TEACHER GUIDE

ADVENTURES in EARTH SCIENCE

    A traditional Earth Science text incl. astronomy which is also available as a series of smaller books:

- Exploration Science (Field Geology and Mapping)
- Riches from the Earth (Minerals, Mining & Energy)
- Changing the Surface (Erosion and Landscapes)
- Rocks - Building the Earth
- Fossils - Life in the Rocks
- A Dangerous Planet (Earth Hazards)
- Through Sea and Sky (Oceanography and Meteorology) Beyond
- Planet Earth (Astronomy)

2018 Digital Edition   ISBN: 978-1-925662-21-4

2018 Print Edition   ISBN: 978-1-925662-20-7

Author: Dr. Peter T. Scott

All illustrations, photographs and videos by the author unless stated
Cover photo: Andrew Scott of AJS Creative

Registration:

Thorpe-Bowker +61 3 8517 8342
email: bowkerlink@thorpe.com.au

No part of this publication may be reproduced, stored in a retrieval system, or transmitted in any form or by any means, electronic, mechanical, photocopying, recording or otherwise, without the prior written permission of the publisher.

© All rights reserved Felix Publishing 2018

About the Author

Dr. Peter T. Scott

Dr. Peter Scott is an award-winning of Earth Science of over forty years' experience in both secondary and tertiary education. He holds a bachelor's degree, two master's degrees and a doctorate, including many years on his own research in locating and correlating coal measures. His experience has been as a classroom teacher and Head of Department but he has also worked at the administrative level as Head of Syllabus and as member of State review panels and has served on advisory bodies for several governments and industry bodies. He has also lectured at university in Science teaching. Apart from his exploration during his own field researches, he has also travelled extensively and has visited many places of interest including Antarctica, the Andes, the Amazon, North Africa, the volcanic islands of the Pacific and Asia, Alaska, Hawaii and the American northwest, northern Europe and remote Australia. His studies and interests have also taken him underground into many limestone and lava caves, coal mines and deep metalliferous mines. The Earth is truly a place of adventure - this book encourages the reader to begin its exploration both at home and abroad.

Table of Contents

Introduction — 1.
1. Why "Adventures" in Earth and Environmental Science? 2. Studying Earth and Environmental Science; 3. Writing Practical Reports; Safety in the Classroom Laboratory; Safety in the Field.

Chapter 1: Exploration Science — 8.
1.1 Introduction to Systems
1.2 Introduction to Scientific Observation: Examination of a Fossil Specimen
1.3 Measurement Exercise: Size of feldspar Crystals.
1.4 Using a Topographical Map

Chapter 2: The Blue Planet — 20.
2.1 Rotation of the Earth
2.2 Tides

Chapter 3: Exploring the Rocky Planet — 24.
3.1 Correlation Exercise
3.2 Mapping Inclined Beds
3.3 Mapping Faulted Beds
3.4 Mapping Folded Beds
3.5 Mapping Igneous Intrusions
3.6 Mapping Unconformities
3.7 Geological History
3.8 Half Life of Radioactive Iodine 131 (Graphing Exercise)
3.9 Size of the Earth

Chapter 4: Rock-forming Minerals — 50.
4.1 Some Common Rock-forming Minerals
4.2 Growing Crystals
4.3 Measuring Specific Gravity
4.4 Geochemistry

Chapter 5: Igneous Rocks - The Beginning — 60.
5.1 Some Common Igneous Rocks
5.2 Igneous Rock Textures and Compositions

Chapter 6: Sedimentary Rocks — 66.
6.1 Some Common Sedimentary Rocks
6.2 Modal Analysis of Sediment

Chapter 7: Metamorphic Rocks — 71.
7.1 Some Common Metamorphic Rocks
Alternative Experiment Some Common Rocks

Chapter 8: Weathering and Erosion — 75.
8.1 Weathering
8.2 Introduction to Stream Table Experiments
8.3 Soil Testing Experiments

Chapter 9: Environments of Weathering and Erosion — 82.
9.1 Karst Simulation
9.2 The Shape of Rivers
9.3 Stereopairs and Landforms

Chapter 10: The Hydrosphere - Waters of the Earth — 88.
10.1 The Properties of Water
10.2 Mapping the Depths
10.3 Water Sampling Techniques

Chapter 11: The Atmosphere - The Air Above — 98.
11.1 Relative Humidity
11.2 The Aneroid Barometer
11.3 Weather Log

Chapter 12: The Biosphere - Life on Earth — 105.
12.1 Introduction to the Biological Microscope
12.2 Examination of a Plant Cell - Onion Epidermal Cells
12.3 Examination of an Animal Cell - Human Cheek Epidermal Cells
12.4 Classification of Organisms
12.5 Examination of Some Common Fossils

Chapter 13: Energy and the Earth — 115.
13.1 Law of Conservation of Energy
13.2 Anaerobic and Aerobic Respiration
13.3 Energy and the Water Cycle

Chapter 14: Energy and the Sea and Sky — 122.
14.1 Radiation and the Inverse Square Law
14.2 Modeling the Greenhouse Effect
14.3 Convection Currents
14.4 Photosynthesis
14.5 Population Growth of a Bacteria Colony (Video Exercise)

APPENDIX A: Risk Assessment of Practical Work and Excursions — 135.

APPENDIX B: Excursion Permission Note — 138.

Introduction

1. Why "Adventures" in Earth and Environmental Science?

The study of the Earth IS an adventure! Studying Earth and Environmental Science involves:

- **Exploration** – whether it be in remote areas or near places of habitation, field work usually involves the breaking of new ground (forgive the pun!). Often in the more remote locations, field work might be over ground which has never been as thoroughly explored as in your study.

- **Exciting Places** - usually outside of the person's usual habitation range and often in very remote places in different parts of the world, in different physical environments such as deep underground (mines and caves), on high mountain ranges and in different climatic conditions (open oceans, hot and cold deserts and jungles).

Photo: Paradise Bay, Antarctica 2011, the remote Argentinian base Almirante Brown

- **Meeting Interesting People** – who are often happy to share their own experiences and culture, not to mention their home, food and local knowledge. One meets a surprising number of people, mostly good people, such as fellow Earth Scientists and students, field workers in the Earth Industries, Government officials; local land-owners and indigenous peoples.

- **Exciting Research** – especially when working on a new project or studying a new location, Earth Science research, as with all scientific studies can be especially motivating (sometimes to the point of happy obsession!). Just like a good detective story, there are things to find out, evidence to gather, step-by-step deductions to be made and finally a conclusion which answers the research question. Looking for gold or gems would be a simple example but it is when the study has a great number of research questions that the field and its companion laboratory work becomes exciting.

- **Studying at many levels** – whether it is a simple prospecting trip or a detailed study of comet movement in the sky over a long time period (NASA leaves this to gifted amateurs!), Earth Science activities can occur at any age and at every level from the school student amateur to the professional Scientist.

Whatever the nature of the study or the level of the student, the Earth is a dynamic and active place with exciting things to see. If Planet Earth is not adventurous enough, there is always the rest of the Universe.

2. Studying Earth and Environmental Science

Earth and Environmental Science cover a wide range of interesting topics, from volcanoes to researching wildlife in Antarctica. The amount of mathematics, often common to Science subjects, has been reduced within the textbook; providing adequate analysis of large amounts of data, but there is still much content to cover in order that a true understanding of the Earth and beyond is obtained.

Because of the amount of content needed to be learned, it is highly recommended that students adopt sound attitudes and practices. These include:

- **Making good, concise summaries of each topic.** Whilst electronic devices such as computers and tablets are excellent for note taking, there is a tendency in modern education for students (and teachers) to upload or cut/paste previously prepared notes. These are good for gathering information as a primary source, but eventually a certain (minimal) form of summary will be required for study purposes and committing a basic set of notes to memory. After all, one must have some information stored in the brain (as well as on Hard Drive) to be able to use when putting into practice what has been learned. The best way is the ancient method of **reading, analysing and extracting the main ideas** and then **writing them down on paper** as a study summary. Doing this on a computer screen has some usefulness but it is not as good for learning as the hand-eye coordination which occurs with physical writing on note paper as a study sheet. Students should organize these sheets so that there is one page per major topic. The use of simple diagrams, charts and lists is an advantage to learning. For long sequences which should be known at all times and not simply retrieved from a data bank on demand, mnemonics are most useful. These may take a simple form of using the first letter of each word in the sequence to make another, simpler word which forms part of a crazy sentence. The crazier, the better. This is then learnt off by rote (try writing it out 100 times!) with the appropriate mental connections made to the real words and sequence.

- **Good study habits** and time management are not natural processes for most students, even those of mature age. A time-management grid or fixed calendar of regular study and assignment preparation could be suggested by the teacher. This would also include the necessary breaks for leisure time and social activities. Such a timetable should be flexible and include such well-known concepts that there should be social/leisure time immediately after classes; study is best done just before sleeping; that a variety of different topics are best done in small amounts rather than a massive amount on the same topic; and that short (written) summaries of the nightly study are best done last. The latter concept does waste a lot of paper but the hand-eye coordination is most useful for memory retention.

- **Completing practical work and assignments on time.** Never leave due work until the last minute. Students should be taught to plan a sequence of preparation e.g. gather notes by library/Internet research and reading, summarizing these notes in a cohesive form, planning the structure of the main work, then writing the assignment in one or several stages so that it is completed well before the due date. Assignments and practical work which occur regularly and often should be started as soon as possible after they have been set. Interruptions due to personal life will occur and if these interfere with the submission of work, extra time should be given by the teacher if they are valid.

- **Noting down ideas**, words etc. not understood should become a regular habit, especially as data recording whilst working is part of the Earth Science method. Difficult questions, new observations and unknown concepts are the source of further research by asking questions. Teachers should encourage their students to always ask questions about what they see and do. There should be a classroom understanding that there is no such thing as either an awkward or silly question. We learn by questioning – even established ideas.

- **Students should work independently and be accountable for their own work.** Although they may have to work and gather data in small groups, students must understand that the final outcome of this work should be their own. It is very easy using electronic media to copy and manipulate other people's original work. Even without computer programs to analyse student work for such plagiarism, it is usually obvious to the trained teacher to see that the level of understanding, terms used and data outlines is not typical of the age nor academic level of the student. Originality, no matter how simple, is the key to a good assignment.

3. Writing Practical Reports

Each school Faculty will have its own view on the recording of any formal practical work completed by the student. It would be hoped that any formal writing, submission and assessment of student practical report will follow a sequence common to that employed later in their studies as post-graduate students or professional Earth Scientists. This sequence is part of the recording phase of the Scientific Method and should be done so that the practical work or experiment can be read and vetted by others and used as a stimulus for further new experimentation and study. A good report or thesis on any research project should follow a logical sequence generate more questions than it answers.

Scientists, like other managers must write down records of their work and present reports. Earth Scientists working for private companies or government agencies write regular reports to their Directors. Research Scientists publish papers (detailed articles for well-known scientific magazines), present seminars and write books about their work. Students write practical reports or theses for schools and universities. Regardless of the type of report, they all follow a similar system: there is an AIM or intent; a detailed account of the METHOD used; some outline of the findings of the research, or their RESULTS; and eventually a CONCLUSION to state what the research proved, how accurate it was and what good it will do.

An example of a sequence of the main components of any written practical reports may be:

1. Giving a **HEADING** and **DATE** (and if required by the teacher, names of co-workers)

2. The **AIM** is written in FUTURE tense e.g. "To ...." There may be one or two aims but never too many. The Aim shows others what you are setting out to show.

3. There may be a need to put in a brief list of **MATERIALS** to show what was used;

4. The **PROCEDURE** is always written in THIRD PERSON (i.e. impersonal e.g. First Person is "I", Second Person is "We" and Third Person does not use personal pronouns at all.) and in PAST TENSE as though the student has already completed the work (because when it is read it **will** have been completed). The Method is best written in point form (some prefer a Flow Diagram but this can be difficult to construct if the method is complex) e.g.

> "1. The specimen was carefully examined with a hand lens;
> 2. Noting its overall shape, an outline was drawn;
> 3. Internal features in the shape were again examined;
> 4. These features were then drawn within the outline;
> 5. Colour and shading was applied; and
> 6. The original specimen's size was compared to that of the sketch to derive a scale in size (e.g. x5)"

5. **OBSERVATIONS & DATA** will include (in third person, past tense) descriptions of ALL observations found using all senses and/or instruments used in detection or any measurement; any data collected; calculations made (some teachers prefer a separate CALCULATIONS section after results); and whenever possible, a DRAWING or SKETCH of any specimen, apparatus or major feature. In Earth Science, drawings of laboratory apparatus are done as in the other sciences such as Chemistry, as 2-dimensional pencil (black ink only if students are good at drawing) using rulers and printed labels in normal ink. Other sketches e.g. of specimen should be done in 3-dimensions in pencil and later coloured pencils and a SCALE of the size (e.g. x1 or x 2 etc.) given. For electronic submissions of work, students are encouraged to use appropriate software for writing the report, doing the artwork and inserting their **own** photographs. If hand-sketched drawings made during the classroom activity are to be included, these should be photographed or scanned **after** completion taking care with colour, size and contrast of the image. Students' own videos may also be included to illustrate important activity but students who have poor computing skills still should be encouraged to submit written reports. The teacher should devise an assessment system in grading reports so that written or electronic submissions are given fair comparisons.

6. **QUESTIONS** if there are specific questions to be answered, it is advisable to also re-write the question and then give its answer.

7. The **CONCLUSION** gives the answer to the aim, any errors encountered and suggestions for improvements. Sometimes, more substantial reports may also have a final section which refers to possible new research.

The first activity (Observation of Pyrite Crystals) is given to you as an example in the way reports will be written in the future. Future activities will have instructions which must be rewritten in the standard manner.

8. Safety in the Classroom Laboratory

Experience has shown that some Secondary and Tertiary students are uncoordinated and socially irresponsible. For this reason, they and rest of the group need to be protected by the application of a set of firm but fair Safety Rules. These should be simple, able to be comprehended by the student and practiced. It is also a good idea to have them posted within the classroom. In time, most groups get into a working pattern and so The Rules become simply standard learning procedures and so a good, working social atmosphere develops. An example of such rules is given on the next page:

SAFETY in the LABORATORY

The following policy is in place to ensure that the laboratories are safe places for students to work in. Respect for self, others and property is first priority. Please ensure that:

1. Safety apparel is to be worn at all times - aprons, good enclosed shoes, eye glasses etc.
2. You do not enter a laboratory without teacher supervision;
3. Unauthorized experiments are strictly forbidden. Variations in procedure must be approved.
4. Food and drink are not consumed in the laboratory;
5. Movement and noise should be kept to a minimum as distractions cause accidents.
6. Items and liquids are not thrown at any time.
7. You do not touch, taste or smell chemicals and minerals only as directed by the teacher.
8. Spillages, breakages and accidents are reported immediately.
9. You know where all safety switches and equipment are located and how they are used.
10. The work area is clean and tidy. When finished, clean apparatus and return with chemicals etc. to appropriate place. Wipe down laboratory bench and wash hands.

In addition, use of electronic equipment such as computers, tablets and cell phones should be appropriate for the learning environment and secondary to the teaching process. Some schools ban the use of cell phones but with a few simple courtesies they can be useful in the classroom.

9. Safety in the Field

This requires more vigilance than in the protected laboratory environment. In the field, away from the classroom, there are other external influences such as:

- the climate, which could be excessive in its heat or cold, windy, wet or subject to sudden change;

- the terrain, which may be steep, rugged, slippery, loose, and full of ravines, caves or mine shafts;

- fast flowing streams or rough seas with unexpected flooding or wave action;

- the vegetation, which may be dense, tangled, thorny and sometimes poisonous; and

- the wildlife, may be dangerous if disturbed or generally a nuisance (such as flies or mosquitos).

Students should also take precautions such as:

- to not wander about and but keep with the group. There is real danger of becoming lost, falling into ravine or mine shaft or other dangers;

- being prepared for the trip, especially about such necessities as:

    CLOTHING - especially adequate footwear, head covering, suitable clothing for the climate and sturdy boots;
    WATER and FOOD as appropriate for the trip;
    SPECIAL SAFETY GEAR (e.g. walking poles);
    INSECT REPELLANT and SUN PROTECTION;
    STUDY ITEMS such as Excursion Guide, notepad and pencil.

- forewarn the teacher about any special needs such as allergies and other problems involving outdoor activities; and

- do not indulge in foolish behaviour such as throwing stones or being a distraction.

Dangerous obstacles should be avoided and only safe, secure tracks should be used. Good navigation is essential and the proposed route and time of return should be left with any local authority such as local police or park rangers as well as with the school authorities. A typical set of Field Safety Rules are:

FIELD RULES

When in the field:

1. Listen to all instructions – especially about specific local hazards;

2. Keep with the group – do not wander;

3. Do not enter bodies of water unless told to do so;

4. Do not enter old mines or industrial workings unless told to do so, then with caution;

5. Do not climb cliffs nor stand under or near unstable rocks;

6. Do not throw any objects, especially hammers nor rocks;

7. Watch your step, especially on slopes and in close vegetation;

8. Wear appropriate field clothing at all times. Be prepared for sudden changes in the weather (rain, cold/heat). Carry a waterproof jacket;

9. Watch out for traffic when on or near roads and railway cuttings;

10. Carry own water and some food;

11. Keep movement and noise to a minimum;

12. Do not use cell phones inappropriately. Headphones are not allowed. Take own care of cameras.

Chapter 1: Exploration Sciences

EXPERIMENT 1.1   Time: One Lesson

INTRODUCTION TO SYSTEMS

<u>**AIM:**</u> To use laboratory equipment and a simple acid-carbonate reaction to demonstrate the differences between a closed and an open system.

<u>**MATERIALS:**</u> Baking soda (bicarb soda i.e. sodium hydrogen carbonate), vinegar (weak acetic acid), 250 ml flask, small test tube which fits into the flask, thermometer, rubber balloon, accurate balance.

<u>**BACKGROUND:**</u>

In an open system both matter and energy can be transferred between the system and the outside environment but in a closed system such as the Earth, matter is retained but energy can be lost or gained.

<u>**PROCEDURE:**</u> (Note how this is written as an example in third person, past tense, and point form)

PART A: An Open System

1. 50 ml of vinegar was added to a 250 ml flask and it and its contents were weighed accurately on the balance and the weight recorded in the data section below. The beaker was placed onto a tray.

2. The temperature of the vinegar was then measured with a thermometer and recorded.

3. About 20 grams of baking soda (or an equivalent carbonate) was measured out and quickly dropped into the flask of vinegar. CAUTION: the reaction may cause the harmless foam to spill out of the flask.

4. When the reaction stopped, the temperature of the contents of the flask was again measured with the thermometer and recorded.

5. The thermometer was removed and the beaker and its contents were reweighed and the weight recorded.

6. The contents of the flask were carefully washed down the sink.

PART B: A Closed System:

1. About 20 g of baking soda was carefully added to a dry flask.

2. A small test tube was half filled with vinegar (assumed to be at the same temperature as before).

EXPERIMENT 1.1 continued

PROCEDURE: Part B continued

4. A balloon's open end was stretched tightly over the opening of the flask so that it made a perfect seal.

5. The entire apparatus was then accurately weighed on the balance.

6. CAREFULLY the palm of the hand was placed over the top of the flask with the balloon still securely in place and the flask was quickly inverted then restored to its upright position and the hand removed from the top to allow freedom of movement of the balloon. The base of the flask was held to see if there was a change in temperature.

7. When the reaction had stopped, the entire apparatus with the balloon still secured was reweighed and the weight was recorded.

8. The balloon carefully removed and the temperature of the contents in the flask were measured.

9. The equipment was carefully washed and returned.

DATA and OBSERVATIONS Part A:

Before Reaction:

Mass of vinegar and beaker = _____ grams

Temperature of vinegar = _____ ^0C

Mass of Baking Soda (or carbonate) = _____ grams

After Reaction:

Describe in words what happened:

Mass of beaker and contents = _____ grams

Describe the reaction between the Baking Soda and the vinegar.

QUESTIONS Part A:

1. Was there any matter lost from the system? If so, what was it?

2. Was there any heat lost or gained from the system? If so, what type of reaction was it (exothermic where heat is lost or endothermic where heat is gained)?

EXPERIMENT 1.1 continued

DATA and OBSERVATIONS: Part B

Mass of apparatus before the reaction   =   grams

Mass of apparatus after the reaction   =   grams

Temperature of the system at end   =

Describe the reaction between the baking soda and the vinegar.

QUESTIONS Part B:

1. Was there any matter lost from the system? If so, what was it?

2. Was there any heat lost or gained from the system? If so, what type of reaction was it (exothermic where heat is lost or endothermic where heat is gained)?

CONCLUSIONS

Considering both reactions, which one was the closed system and which one was the open system?

Why was one of these procedures considered a closed system?

Was the heat change generally the same in both procedures? If not explain any differences between the heat change of both procedures.

Comment on this experiment as a model to show open and closed systems. What errors could have occurred?

EXPERIMENT 1.2

Time: One Lesson

INTRODUCTION TO SCIENTIFIC OBSERVATION: EXAMINATION of a FOSSIL SPECIMEN

AIM: To use a hand lens or binocular microscope to observe, sketch and describe pyrite crystals

MATERIALS: Fossil leaf or other specimen   hand lenses or binocular microscopes   coloured pencils

BACKGROUND:

Fossils and other specimens must be examined carefully for any detailed features which would aid in their identification e.g. vein structure in leaves. CAREFUL OBSERVATION is needed to distinguish one species from another. Sketching the specimen to scale (usually about x3 or x4) is a good way to record any special features for later comparison with other specimen.

PROCEDURE: (NOTE: THIRD PERSON, PAST TENSE, and POINT FORM)

PART A: SETTING UP THE MICROSCOPE (if binocular/dissection microscopes are available)

1. The microscope was carefully removed from its box and the box put aside.

2. The large black knob on the upright rod was loosened and the whole body of the microscope was moved up to near the top of the rod.

3. The eyepieces were adjusted so that one complete circle was seen.

4. The specimen was placed in the centre of the stage and the eyepiece housing was moved down to near the specimen using the black focusing knobs.

5. Looking down the eyepieces the focusing knob was adjusted UPWARD until a clear view was seen.

PART B: OBSERVING DETAIL

6. The overall shape of the specimen was carefully examined and compared with common shapes (e.g. cubes, rectangles, triangles, ovals etc.).

7. Details of COLOUR, SHADING and INTERNAL SHAPES (i.e. shapes within the specimen and how they are placed together) were observed as were any INTERNAL DETAILS such as lines, patterns etc....any regular patterns.

EXPERIMENT 1.2 continued

PROCEDURE: continued,

PART C: SKETCHING THE SPECIMEN

8. A circle was drawn (about one third of a page) to represent the field of view of the microscope and a rule was placed under the microscope to measure the actual field of view (e.g. about 10 mm). This was written near the circle to give scale.

9. Thoughts about what was seen (shape, colour, position of parts etc.) were revised and sketched using very light pencil;

10. Detail was re-observed and included in the sketch and the outline was re-done with heavier pencil (or black ink if competent).

DATA and OBSERVATIONS:

Copy a circle into which the sketch can be made (DO NOT DRAW IN THIS BOOK):

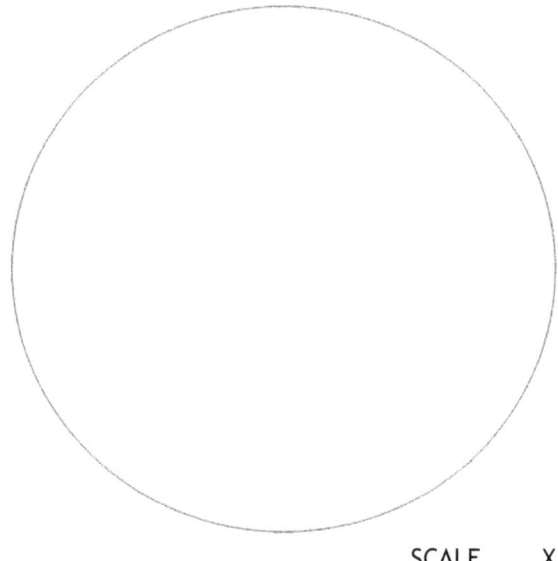

SCALE   X

Do a detailed sketch in colour and add labels for distinctive features. Also give the scale. (HINT: Draw the outline of the specimen without the aid of the microscope or hand lens so that it almost fills the circle. Look through the microscope and find the detail. From memory, return to the sketch and draw in the detail. Add shading and colour last. Compare the size of the sketch to the original to get scale.

Describe the main features observed e.g. overall shape, internal features and patterns etc.

EXPERIMENT 1.2 continued

CONCLUSIONS

1. What are the main distinguishing features of Pyrite?

2. Comment on factors which might limit the accuracy of observation.

3. Why is it important to have an accurate scale in drawing?

4. Other general conclusions (if required)

(NOTE: Unlike the last two activities, students are now required to rewrite their reports from this document as aim, method, results, and conclusion using third person, past tense, and point form. materials and background may not be required in your report).

EXPERIMENT 1.3   Time: One Lesson

MEASUREMENT EXERCISE: SIZE of FELDSPAR CRYSTALS

Use a photomicrograph (photo taken through a geological microscope) to estimate the size of feldspar crystals) and determine any relationship between size of crystals and depth of their formation.

MATERIALS: Three photomicrographs (each x 25 magnification) of separate specimens of basalt rock, rulers, and calculators as required.

BACKGROUND:

Three photomicrographs (each x 25 magnification) taken of thin-sections of separate specimen of a basalt rock found at different depth. Each photomicrograph has been made by cutting a piece of the basalt to a thin slice which is then glued to a glass slide and ground down until transparent. It is then viewed in polarized light and photographed

PROCEDURE:

For each photomicrograph measure a good number (say 20) of the lengths of the feldspar crystals (seen as long shapes) in each of the three photographs. The scale of each photo has been made up to the real size of the crystals (i.e. what you measure with your ruler is the correct size)

Plot your data in a Table for each of the photographs relating sizes and depths:

e.g.

Photo 1 (Depth =)	Photo 2 (Depth =)	Photo 3 (Depth =)
measurement 1 mm measurement 2 etc. etc. (20 measurements) Average = mm	measurement 1 mm measurement 2 etc. etc. Average = mm	measurement 1 mm measurement 2 etc. etc. Average = mm

Use your knowledge of mathematics to find an average size of the crystals for each photomicrograph and estimate the error of measurement for your final value.

Make a sketch of one of the photomicrographs as a representation of those given. Provide a scale for your sketch.

EXPERIMENT 1.3 continued

Photomicrograph 1 (Depth 2 metres below surface)

Photomicrograph 2. (10 metres below surface)

Photomicrograph 3 (20 metres below surface)

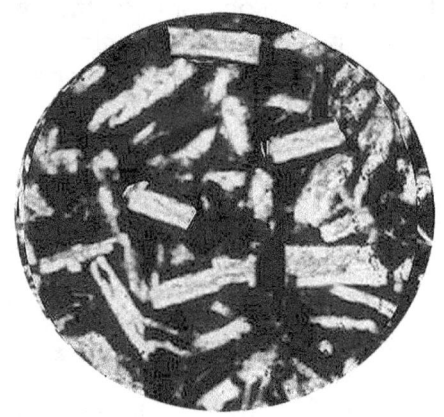

CONCLUSIONS:

1. What was the average size of the crystals for each of the depths and their error of measurement?

2. What type of error of measurement is this?

3. Is there a relationship between crystal size and depth of their formation? If so, what is it? Can you express this relationship as a mathematical expression?

4. Are your measurements made on the photographs a true representation of the sizes of the crystals? Why? How can you improve on the method of **randomly** sampling which crystals are to be measured?

5. What are some other conclusions about the size of crystals or method used?

6. How could you further test your hypothesis about size of crystals and depth?

What are feldspar crystals? How do they form in rocks such as basalts? Where would very large crystals of feldspar (say a metre long) form? What use is feldspar in daily life?

| EXPERIMENT 1.4 | One or two Lessons |

USING A TOPOGRAPHICAL MAP

AIM: To use a topographical map to locate and refer to objects or features, measure distance and draw a simple cross-section to scale.

MATERIALS: Pencils, rulers, grid/graph paper, paper, string

BACKGROUND:

Topographical maps are still most useful despite the dependence on GPS devices. They can be used in obtaining a general idea of a research area and some of its features. Electronic devices can sometimes fail in some remote areas. Aerial photos showing the real (as opposed to mapped) features are also very useful.

PROCEDURE:

(a) General Features of a Topographic Map.

Look at the general features of the topographical map on the next page:

QUESTIONS:

1. What is the scale of this map? What does this mean in reality?

2. What are the geographical coordinates (i.e. Latitude & Longitude of this area?

3. Use an atlas (printed or electronic) to find where this place is located.

4. What is the magnetic variation of Utopia Point? (as degrees East)

5. List the grid numbers which represent:
    a. the eastings
    b. the northings

6. Use the scale and a ruler to measure the length of the jetty at Utopia Point.

7. Use a piece of string to estimate the length of the coastline shown on the Map.

8. Within which grid square is Hospital Point?

9. What objects are located at grid reference:
    a. 393583
    b. 428590

10. What is the grid reference for:
    a. the end of the jetty at Utopia Point
    b. the lone hill (118 m high) in the far northeastern part of the map

11. What is the contour interval of this map?

12. What is the feature between grids 413576 and 416572?

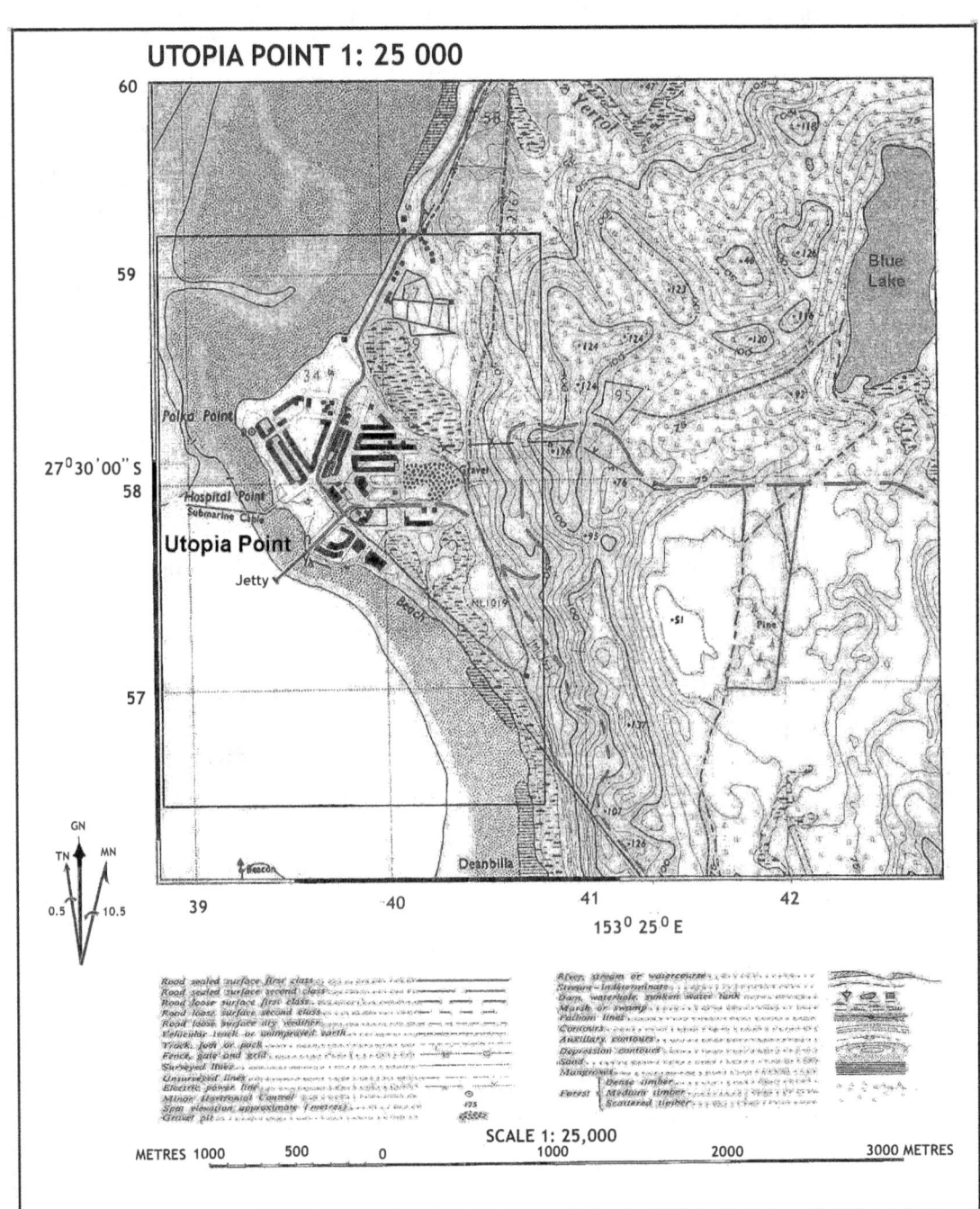

EXPERIMENT 1.4 continued

PROCEDURE (continued):

Drawing a cross-section.

Whilst there are computer programs which will draw map cross-sections, it is still useful to know how it is done and to understand the concepts of vertical exaggeration, gradients and the basic shapes of land features.

Consider the following large scale part of a map:

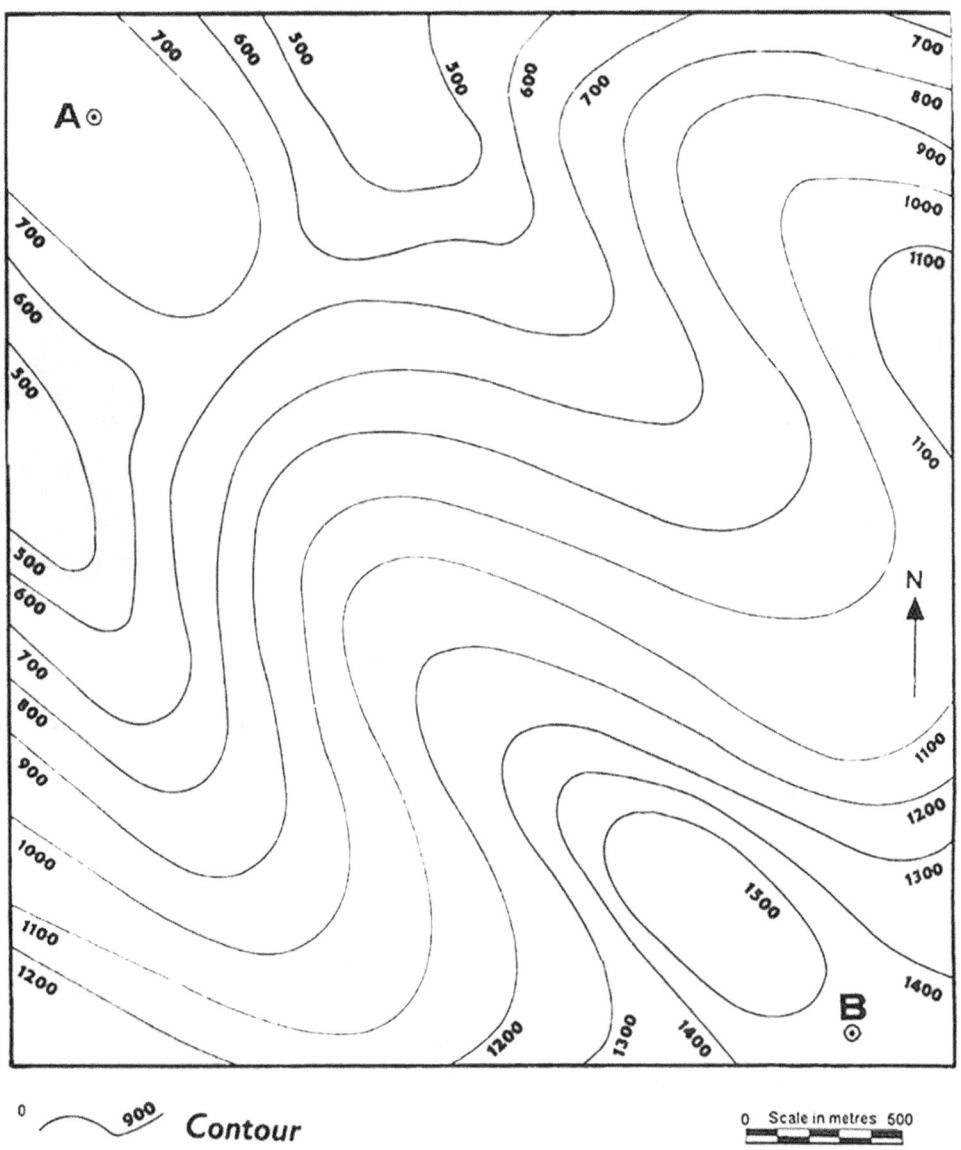

EXPERIMENT 1.4 continued

PROCEDURE:

1. Copy the cross-section box below so that the horizontal distance A-B is the same as on the map(do not draw in this book);

2. Look at the heights as shown on the contour intervals. These give the maximum and minimum heights to use on the vertical axis of the cross-section box;

3. Using the same scale as on the map (i.e. vertical scale now equals horizontal scale), construct the vertical sides of the cross-section box down to sea level (0 metres);

4. Us a straight-edged piece of paper along A-B on the map to mark of where the contour lines cut line A-B. Also mark on the paper the values of the heights;

5. Transfer these marks and values onto the top A-B line of the cross-section box;

6. Going from A to B, draw faint construction lines down from each mark on A-B down to its appropriate height above sea level on the vertical scale of the box. Mark the end of each construction line with a dot;

7. Using some estimation and rounding to represent a land surface, join up the dots to draw in the topography of the cross-section.

QUESTIONS:

1. What feature lies between spot heights A and B?

2. When drawing this cross-section, what allowances to reality must the drawer make?

3. What is the gradient of the steepest slope along section A-B?

4. If this is open grassland country and in good weather, what would be an estimate of the time it would take to walk (unburdened) from A to B?

CONCLUSIONS:

1. What would be the (a) advantages and (b) disadvantages of using a paper topographical map over that of an electronic, hand-held Mapping Ap. or computer screen map?

2. Why would an Earth Scientist use such a map? (consider pre- and post- field work activities)

3. What are some general conclusions about this mapping exercise?

Chapter 2: The Blue Planet

EXPERIMENT 2.1 One lesson or school or home time

THE ROTATION OF THE EARTH

AIM: To find out how fast the Earth rotates by experiment.

MATERIALS: metre rule   very large protractor
shadow stick (a metre rule or a two metre long stick - the longer the better)

BACKGROUND: This method and its geometry were well-known to the Ancient Greeks and the principle was used in the construction of sundials.

PROCEDURE: (WARNING: Never look directly at the Sun)

1. Place a shadow stick of known height in the centre of an open area exposed to direct sunlight ensuring that the stick is vertical and placed on a large sheet of paper or on a flat surface which can marked or on which a shadow can be measured. Mark or note the first position of the base of the stick and along the line of the first shadow (or leave the shadow stick in place).

2. Measure the length of the first shadow.

3. Return in a few minutes (say every 15 min.) and note the position of the end of the new shadow.

4. Measure the length of this shadow.

5. Measure the angle (θ_1 to θ_3 etc.) between the line subtended from one shadow to the (e.g. from position 1 and the line now formed from the stick by the new shadow at position 2) as accurately as possible using a large protractor.

6. Repeat steps 4 and 5 for each new shadow.

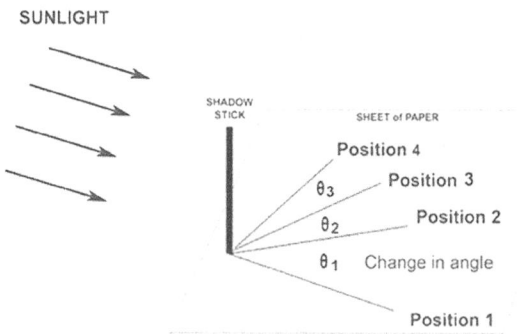

7. (GRAPH 1.) Graph the length of each shadow against the actual time (e.g. 10.00 am, 10.15, 10.30, 10.45, 11.00, 11.15, 11.30. 11.45, 12.00, 12.15 pm, 12.30, 12.45, 1.00 etc. depending upon time available in class or at home).

EXPERIMENT 2.1 continued

PROCEDURE: continued

8. (Graph 2.) Graph the cumulative angle (i.e. the continued addition of all angles from the start) against the actual times (as in step 7) using a time scale of 12 hours.

9. If measurements are extremely accurate, extrapolate the graph shape of graph 2. To correspond to 12 hours and find the corresponding angle. This should give the rotation of the Earth for 12 hours so multiply by 2 and find how long does it take for one complete rotation (note: this must be done very accurately and results often give large errors!)

    Alternatively: One could also redraw the graph, extrapolating its shape to find the number of hours through which the shadow would pass as it turns through 360^0

DATA and OBSERVATIONS:

Record each of:
- Length of the shadow stick (centimetres):
- Length of the shadows in each position at each time:
- Each angle between each shadow:
- Time between measurements:

CONCLUSIONS:

1. Describe the shape of graph 1.
2. What was the time for the shortest shadow (solar noon) from this graph?
3. Was this value exactly 12.00 noon? If not why was it different?
4. What was the shape of graph 2? What does it say about the rotation of the Earth?
5. If the shape of graph 2. was extrapolated, what was the angle or time for one rotation of the Earth. Calculate the percentage error for this calculation.
6. How could this measurement (step 9) be made more accurate?

Research (Optional)

Use the Internet to find out about

1. Sidereal time and how Earth time is measured today.
2. The instruments used by Tycho Brahe.

EXPERIMENT 2.2

One lesson

TIDES

AIM: To graph some tidal data and explain the type of tidal current represented

MATERIALS: Graph paper if required, tidal data and map (supplied), pencils

BACKGROUND:

Tides are the changes in the local levels of the sea due to the gravitational pull of the Moon and to a lesser extent the Sun, on the waters of the Earth.

PROCEDURE:

Study the following tide chart data:

DAYS	TIME 24 hour	HEIGHT metres	TIME 24 hour	HEIGHT metres	TIME 24 hour	HEIGHT metres	TIME 24 hour	HEIGHT metres	MOON PHASE
Mon 4 Nov	0229	1.44	0800	0.78	1502	2.01	2215	0.77	Last Quarter
Tue 5 Nov	0355	1.45	0919	0.86	1611	1.97	2315	0.71	
Wed 6 Nov	0510	1.56	1042	0.85	1715	1.9			
Thurs 7 Nov	0006	0.63	0605	1.70	1151	0.78	1808	2.02	
Fri 8 Nov	0049	0.54	0650	1.86	1245	0.70	1853	2.05	
Sat 9 Nov	0127	0.45	0730	2.00	1333	0.63	1933	2.07	
Sun 10 Nov	0202	0.38	0806	2.12	1417	0.57	2010	2.07	
Mon 11 Nov	0235	0.33	0843	2.22	1459	0.53	2045	2.04	
Tue 12 Nov	0307	0.30	0919	2.30	1540	0.52	2119	2.00	Full Moon
Predictive Data from the Australian Bureau of Meteorology for the Brisbane bar between 4th and 12th November 2019 Note that time uses the 24 hour system i.e. 1 pm = 1300, 2pm = 1400 etc. Add 12 to times after noon and heights are taken relative to Mean Lower Low Water (MLLW).										

Using a large sheet of graph paper, draw up the axes so that the horizontal axis shows the days and times as 4 divisions for each day i.e. 0600 (6 am), 1200 (12 noon), 1800 (6 pm) and 2400 (midnight) and the vertical axis shows the heights in metres above mean sea level from 0 metres to 3.0 metres;

Carefully plot dots for the positions shown in the chart;

Carefully draw curves of best fit through these dots;

Label the graph with a heading and names and units of each axis.

(For electronic reports, the finished graph can be scanned and inserted or a graph App may be used if appropriate)

EXPERIMENT 2.2 continued

DATA and OBSERVATIONS:

Insert the graph here e.g.:

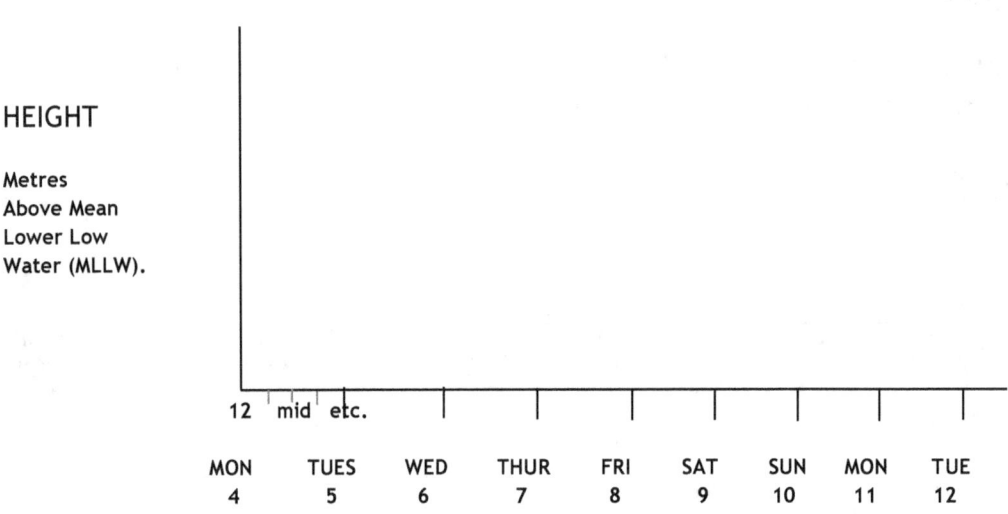

Describe the shape of the graph and note when the highest and lowest tides occur.

CONCLUSIONS:

What does the graph show about tides?

When did the highest and lowest tides occur? Why?

What other factors would affect the time of the various tides in this location?

Why are a knowledge of tides useful?

What type of tidal pattern does this graph show?

RESEARCH (Optional at the teacher's discretion):
Locate a local time chart for the current month. Does it agree with findings from this experiment?

Chapter 3: Exploring the Rocky Planet

EXPERIMENT 3.1 One lesson

CORRELATION EXERCISE

<u>AIM:</u> To use fossils to correlate sedimentary beds at different locations.

<u>MATERIALS:</u> Coloured pencils   ruler   glue   sheets of blank paper

<u>BACKGROUND:</u>

Correlation is the comparison of two or more sequences of rock to see if they had been formed at the same time. This relies on the several principles of stratigraphy and the Principle of Uniformitarianism assuming that the processes which formed these sequences were formed under the same sedimentary processes observed today. Correlation can be done by comparing the fossils found within sedimentary rock sequences by looking at the separate rock units at different localities. This can be done on an international scale or at a more local level.

Correlation at a local level is used when geologists map how rocks extend below the ground. To support their findings about surface rocks, geologists will employ drilling teams to drill below the surface. The drills are hollow lengths of pipe which are screwed into the ground by the motor of the drill rig. These lengths of pipe are then brought to the surface, uncoupled from the drill rig and opened to obtain the drill cores which can then be stored in a drill core library.

<u>PROCEDURE:</u>

1. Carefully examine all of the rock units containing fossils (named as **F1, F2** etc.) within each of the columns given on the next page. These columns represent rocks at four different localities which may have had different formations:

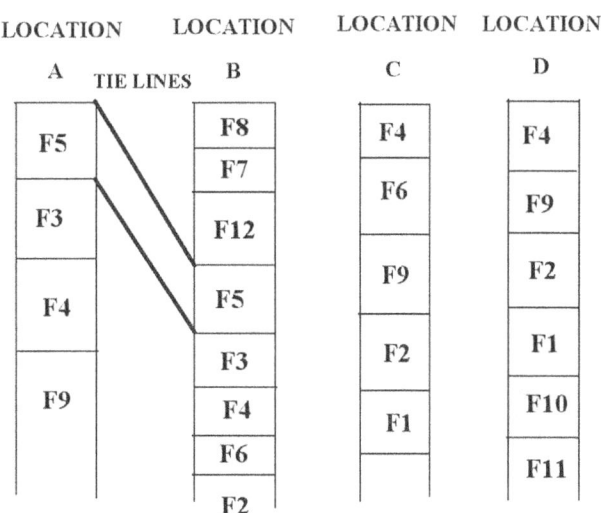

Stratigraphic columns for four localities.
Codes F1-F12 represents fossils and the thickness of the beds which contain them.

EXPERIMENT 3.1 continued

PROCEDURE: continued

2. Copy or photocopy this diagram and then join beds which have the same fossils by using tie lines to the tops and bottoms of these beds (one has been done as an example).

3. Looking at the overall connection, determine which are the oldest and youngest beds using the Law Of Superposition (younger beds are placed on top of older beds).

4. List all of the beds in order from **oldest** (which was deposited first) to **youngest.**

DATA and OBSERVATIONS:

Copy, cut and paste the stratigraphic columns and their tie lines to show which beds contain the same fossils and are therefore of the same age. If required scan or photograph this arrangement.

List the sequence of the beds from oldest to youngest.

CONCLUSIONS:

1. What errors could occur in using this method to determine relative age of beds?
2. How important is the accuracy of identification of the fossil specimen in this process?
3. Give an hypothesis for why Fossil **F6** is missing from columns A and **D.**
4. This exercise has fossils in all rock units. This is not common as fossils are very difficult to find. Give some reasons why fossils may not be found in rock units at all.

EXPERIMENT 3.2   One or two lessons

MAPPING INCLINED BEDS

AIM: To draw a cross-section of inclined beds down to a suitable depth.

MATERIALS: Pencils, rulers, grid/graph paper, paper, protractors

BACKGROUND:

In Chapter One, students were shown how to make topographical cross-sections. Geologists can use a variety of techniques to find out what is happening to rocks and geological structures below the surface. The most basic and traditional way is by measuring the orientation of beds on the surface (dip and strike) and then using simple geometrical concepts to construct scaled drawings of how these beds may look like below the surface. Usually, the first part of these investigations involve surface mapping and drawing of topographical cross-sections. In the following exercises for simplicity, the surface will be considered to be flat.

PROCEDURE:

PART A. Mapping Dipping Beds In the Dip Direction where normally horizontal beds (symbol: +) have been tilted by Earth forces to a measureable angle of dip and the geologist has obtained data along the dip direction (at 90^0 to the strike which is North-South here).

Look at the plan view (i.e. taken looking down onto the surface) of Map 1. on the next page.

1. Draw a topographical cross-section rectangle where the top is equal to the length A-B and represents the surface. Complete the rectangle so that it is about 6 cm deep.

2. Place the edge of a spare paper sheet along the map from A – B and mark off the locations of A, B and the boundaries of each bed. Also mark the angles of dip for each bed and the type of rock represented by the symbols.

3. Transfer this data to the drawn cross-section rectangle by placing the same paper edge between A and B and marking off the bed boundaries (including their dip).

4. Using a protractor along the top of the rectangle (= the land surface), draw in lines at the appropriate dip (say 40^0) in the dip direction (here, towards A) to represent he bedding planes.

5. Complete the cross-section by adding a small amount of rock symbol (as a representative of the whole bed) and the angle of dip for each bed.

EXPERIMENT 3.2 continued

PROCEDURE: continued

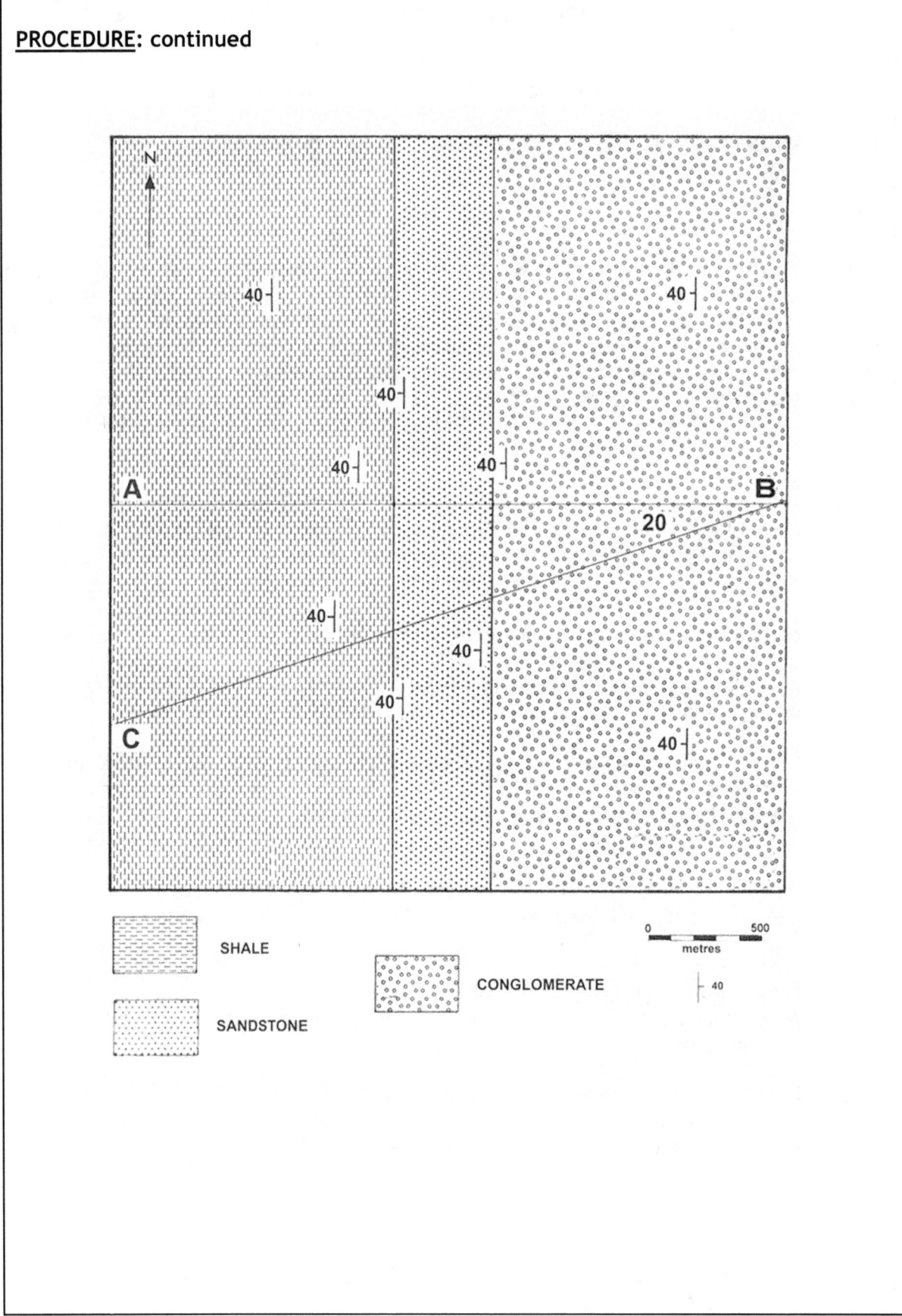

EXPERIMENT 3.2 continued

PROCEDURE: continued

PART B. Mapping Dipping Beds Not In the Dip Direction (Apparent Dip)

Here the geologist has taken data whilst walking obliquely across the dip direction in a direction of 20° (the direction angle) to it. Any angle of dip measurement will not be the true dip but an apparent dip if the geologist does not allow for it. Usually the geologist will find a bedding plane and compensate for this apparent dip by looking at the surface of the plane and deciding where the true dip direction is located and then measuring the true dip accordingly (one trick is to dribble a stream of water onto the plane and it will run downhill in the true dip direction). In this exercise, the geologist wishes to know what the beds will look like below the surface along the traverse B-C.

1. Draw a topographical cross-section rectangle where the top is equal to the length B–C as before to represent the surface. Complete the rectangle so that it is about 6 cm deep.

2. Place the edge of a spare paper sheet along the map from B - C and mark off the locations of A, B and the boundaries of each bed. Also mark the true angles of dip for each bed and the type of rock represented by the symbols.

3. Use the table which converts true dip to apparent dip (Table 16.1 of text) using the direction angle (20°) and the true dip (40°).

True Dip	Direction angle - angle between True Dip and Traverse															
	10°	15°	20°	25°	30°	35°	40°	45°	50°	55°	60°	65°	70°	75°	80°	85°
	Apparent dip															
10°	10°	10°	9°	9°	9°	8°	8°	7°	6°	6°	5°	4°	3°	3°	2°	1°
15°	15°	14°	14°	14°	13°	12°	12°	10°	10°	9°	8°	6°	5°	4°	3°	1°
20°	20°	19°	19°	18°	18°	17°	16°	14°	13°	12°	10°	9°	7°	5°	4°	2°
25°	25°	24°	24°	23°	22°	21°	20°	18°	17°	15°	13°	11°	9°	7°	5°	2°
30°	30°	29°	28°	28°	27°	25°	24°	22°	20°	18°	16°	14°	11°	9°	6°	3°
35°	35°	34°	33°	32°	31°	30°	28°	26°	24°	22°	19°	16°	13°	10°	7°	4°
40°	40°	39°	38°	37°	36°	35°	33°	31°	28°	26°	23°	20°	16°	12°	8°	4°
45°	45°	44°	43°	42°	41°	39°	37°	35°	33°	30°	27°	23°	19°	15°	10°	5°
50°	50°	49°	48°	47°	46°	44°	42°	40°	37°	34°	31°	27°	22°	17°	12°	6°
55°	55°	54°	53°	52°	51°	49°	48°	45°	43°	39°	36°	31°	26°	20°	14°	7°
60°	60°	59°	58°	58°	56°	55°	53°	51°	48°	45°	41°	36°	30°	24°	17°	9°
65°	65°	64°	64°	63°	62°	60°	59°	57°	54°	51°	46°	42°	36°	29°	20°	11°
70°	70°	69°	69°	69°	68°	67°	65°	63°	60°	58°	54°	49°	43°	35°	25°	13°
75°	75°	74°	74°	74°	73°	72°	71°	69°	67°	65°	62°	58°	52°	44°	33°	18°
80°	80°	80°	79°	79°	78°	78°	77°	76°	75°	73°	71°	67°	63°	56°	45°	26°
85°	85°	85°	85°	84°	84°	84°	83°	83°	82°	81°	80°	78°	76°	71°	63°	45°

EXPERIMENT 3.2 continued

PROCEDURE: continued

5. Using a protractor along the top of the rectangle (= the land surface), draw in lines at the appropriate apparent dip in the new direction dip (here, towards C) to represent he bedding planes.

6. Complete the cross-section by adding a small amount of rock symbol (as a representative of the whole bed) and the angle of dip for each bed.

DATA and OBSERVATIONS:

Include cross-sections for Parts A and B here showing the dipping beds along these traverses.

CONCLUSIONS:

1. Why would a geologist not walk along the dip direction to measure true angle of dip?

2. What other geological or environmental situations cause difficulty in measuring the true dip of beds?

3. Why is dip direction a better parameter than strike when discussing the orientation of beds?

4. In Part A, at what depth would a drill core meet the top of the conglomerate if the drill was started on the surface along the traverse at the contact of the shale and sandstone (Hint: use the cross-section and its scale)?

Research (Optional)

Use the textbook or the Internet to revise how angle of dip and the dip direction is measured in the field.

EXPERIMENT 3.3   One lesson

MAPPING FAULTED BEDS

AIM: To draw a cross-section across a fault line and then measure the throw and heave of the fault.

MATERIALS: Pencils, rulers, grid/graph paper, paper, protractors

BACKGROUND:

There are several types of faults where the subsurface beds have been shifted up, down or horizontally along a fault line. Walking along a traverse, a geologist may suspect that there is a subsurface fault if there has been a sudden change in the beds such as one bed recurring further along the traverse. Subsurface mapping will show the parameters of the fault such as its throw (the vertical displacement) and the heave (the horizontal displacement) if they cannot be seen on the surface.

PROCEDURE:

Look at the plan view (i.e. taken looking down onto the surface) on the next page.

1. Draw a topographical cross-section rectangle where the top is equal to the length A-B and represents the surface. Using an appropriate scale (say 1 cm = 100 metres depth), complete the rectangle so that it represents a depth of 600 metres.

2. Place the edge of a spare paper sheet along the map from A – B and mark off the locations of A, B and the boundaries of each bed. Also mark the angles of dip for each bed and the type of rock represented by the symbols and the position and dip of the fault line.

3. Transfer this data to the drawn cross-section rectangle by placing the same paper edge between A and B and marking off the bed boundaries (including their dip) and the position of the fault line.

4. Use a protractor to draw a line right through the cross-section at the dip ($70°$) of the fault line (as one does not know the depth of the fault and so assume that it goes beyond 600 m deep).

5. Using the protractor along the top of the rectangle (= the land surface), now draw in lines at the appropriate dip (say $40°$) in the dip direction (here, towards A) to represent he bedding planes but do not run the lines through the fault line previously drawn.

6. Now the base of any bed on the other side of the fault line must be estimated by finding its thickness (distance at $90°$ from top to bottom bedding planes for that bed) on the other side of the fault.

7. Complete the cross-section by adding any unknown bedding plane (use dotted lines) and then adding a small amount of rock symbol (as a representative of the whole bed) and the angle of dip for each bed.

EXPERIMENT 3.3

PROCEDURE: continued

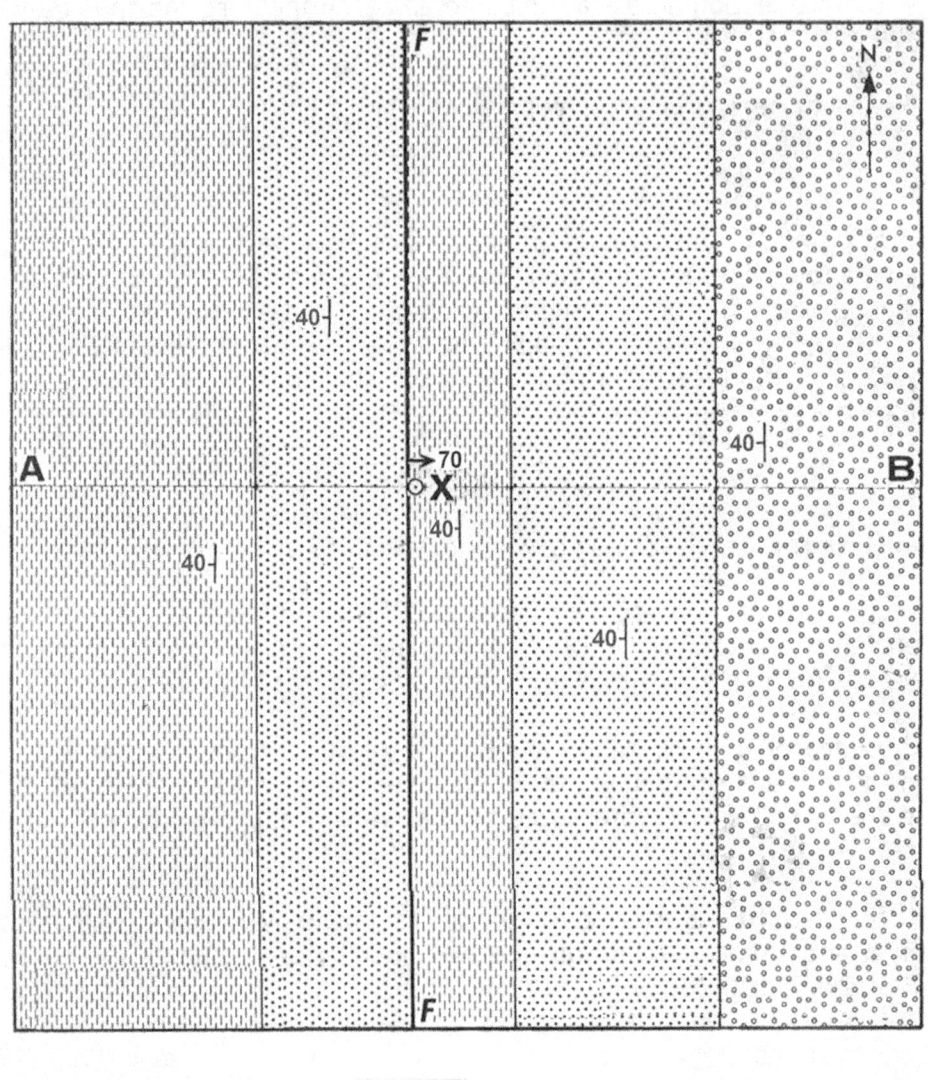

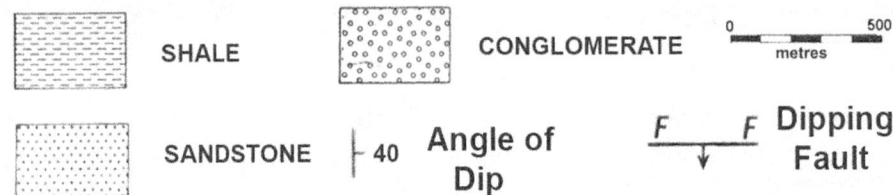

EXPERIMENT 3.3 continued

DATA and OBSERVATIONS:

Draw the cross-section for map showing the dipping beds along the traverse and the subsurface fault line.

CONCLUSIONS:

1. What type of fault is this? Which side of the fault has moved up or down?

2. Which side of the fault is the (a) hanging wall (b) the footwall?

3. What is the (a) heave and (b) throw of the fault?

4. What type of Earth forces would cause such a fault?

5. If a drill was put down vertically at X, at what depth would it strike the top of the conglomerate bed?

Research (Optional)

Use the textbook or the Internet to revise the different types of faults which undergo vertical movement.

EXPERIMENT 3.4   One lesson

MAPPING FOLDED BEDS

AIM: To draw a cross-section across folded beds.

MATERIALS: Pencils, rulers, grid/graph paper, paper, protractors

BACKGROUND:
Beds are usually folded by compressional forces. They can be simple upward folds (anticlines), downward folds (synclines) or a complex series of folds with equally angled limbs or sides (symmetrical) or with limbs of different angles of dip (asymmetrical).

PROCEDURE:

Look at the plan view (i.e. taken looking down onto the surface) on the next page.

1. Draw a topographical cross-section rectangle where the top is equal to the length A-B and represents the surface. Using an appropriate scale (say 1 cm = 100 metres depth), complete the rectangle so that it represents a depth of 600 metres.

2. Place the edge of a spare paper sheet along the map from A – B and mark off the locations of A, B and the boundaries of each bed. Also mark the angles of dip for each bed and the type of rock represented by the symbols and the position of each fold axis (for simplicity, assume here that these are vertical).

3. Transfer this data to the drawn cross-section rectangle by placing the same paper edge between A and B and marking off the bed boundaries (including their dip) and the position of the fold axes.

4. Draw vertical dotted lines vertically right through the cross-section at each fold axis (as one does not know the depth of the fold and so assume that it goes beyond 600 m deep).

5. Using the protractor along the top of the rectangle (= the land surface), draw in lines at the appropriate dip in each of the different the dip directions to represent he bedding planes which have been folded and so dip in these directions. Do not run the lines through the fold axes lines previously drawn.

6. Now the base of any bed on the other side of the fault line may have to be estimated by finding its thickness (distance at $90°$ from top to bottom bedding planes for that bed) on the other side of a fold axis.

7. Complete the cross-section by adding any unknown bedding plane (use dotted lines) and then adding a small amount of rock symbol (as a representative of the whole bed but care with orientation of the symbol) and the angle of dip for each bed.

EXPERIMENT 3.4

PROCEDURE: continued

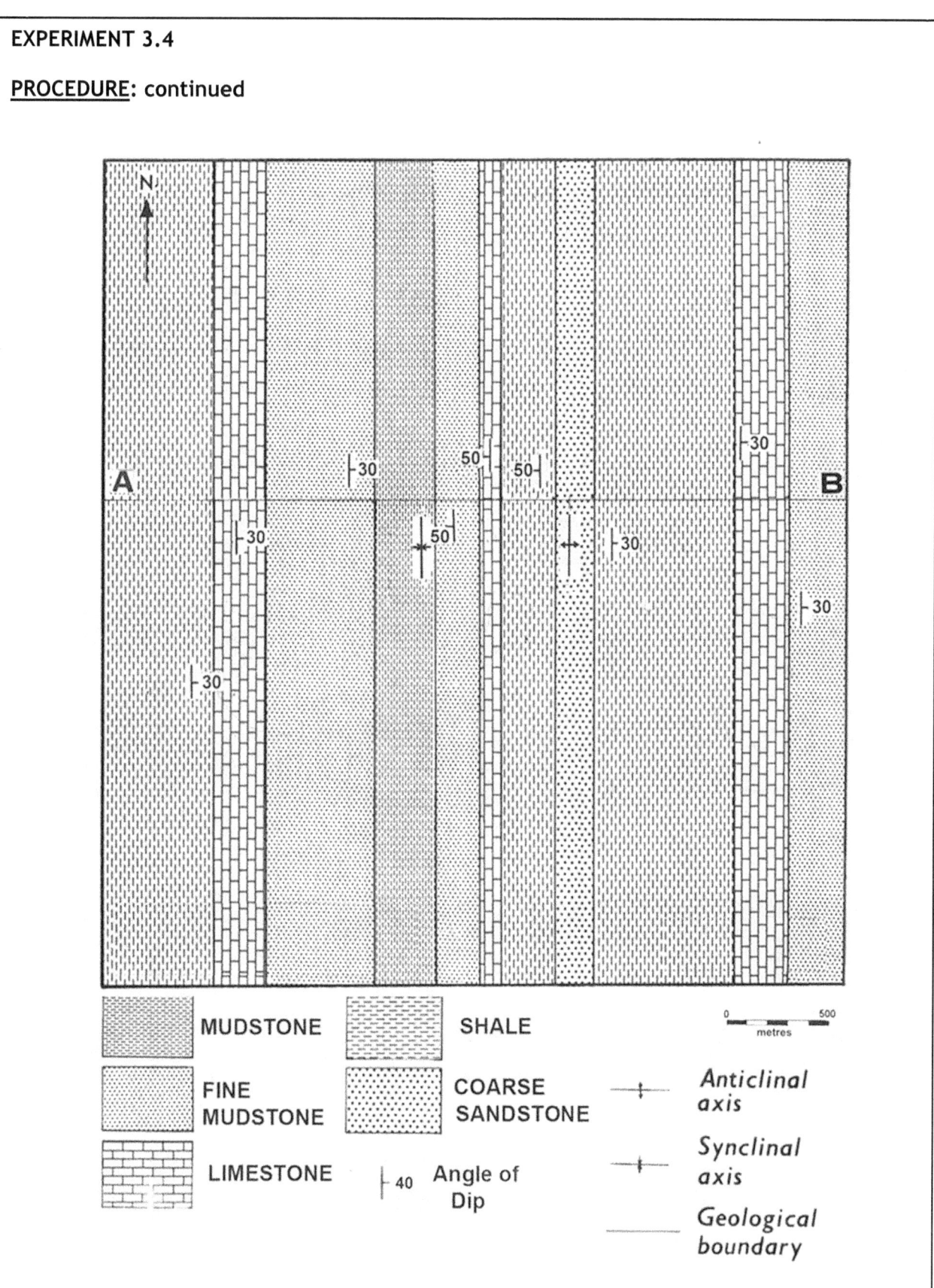

EXPERIMENT 3.4 continued

DATA and OBSERVATIONS:

Draw the cross-section for map showing the folding beds on each side of their axes.

CONCLUSIONS:

1. What type of fold is (a) closer to A and (b) closer to B?

2. What type of Earth forces would cause such folding?

3. Which sedimentary beds were deposited (a) first and (b) last?

4. When did the folding occur in relationship to the deposition of the beds?

Research (Optional)

Use the textbook or the Internet to revise the different types of folding.

EXPERIMENT 3.5

One lesson

MAPPING IGNEOUS INTRUSIONS

AIM: To draw a cross-section across a region has been intruded by several igneous bodies.

MATERIALS: Pencils, rulers, grid/graph paper, paper, protractors

BACKGROUND:

As the name suggests, molten rock (magma) may come up through the exiting layers of rock (called the country rock) in several different forms. These intrusions may be small, thin tabular intrusions (dykes) which come up vertically along joints, cutting across existing rock layers or squeezed in between layers and are parallel to them (sills). They may also be extremely massive and come up vertically as rounded masses (as stocks) or even larger dome-shaped masses over many tens of kilometres across (as batholiths).

PROCEDURE:

Look at the plan view (i.e. taken looking down onto the surface) on the next page.

1. Draw a topographical cross-section rectangle where the top is equal to the length A-B and represents the surface. Using an appropriate scale (say 1 cm = 100 metres depth), complete the rectangle so that it represents a depth of 600 metres

2. Place the edge of a spare paper sheet along the map from A – B and mark off the locations of A, B and the boundaries of each bed. Also mark any angles of dip for each bed and the type of rock represented by the symbols and the position of each of the intrusions where they cut the surface. For simplicity, assume here that these are vertical with dykes being thin with parallel sides and stocks and batholiths coming up from a larger base.

3. Transfer this data to the drawn cross-section rectangle by placing the same paper edge between A and B and marking off the bed boundaries (including their dip) and the position of the edges of the intrusions.

4. Draw lines down through the cross-section for any dykes. If no dip is given for them assume that they are vertical, otherwise draw in the parallel sides of the dyke with the appropriate dip.

5. Draw in dotted lines (of uncertainty) down and spreading out from the surface edges of any stock or batholith to represent that they are a much larger mass below ground. Where these subsurface edges of these intrusions will not be known unless drill cores are taken.

6. Using the protractor along the top of the rectangle (= the land surface), draw in lines at the appropriate dip in each of the different the dip directions to represent any bedding planes which have been tilted and in their dip directions. Lines for tilted sedimentary beds can be carried through to the other sides of the intrusions.

7. Complete the cross-section by adding a small amount of rock symbol (as a representative of the whole bed but care with orientation of the symbol) and the angle of dip for each bed.

EXPERIMENT 3.5 continued

PROCEDURE: continued

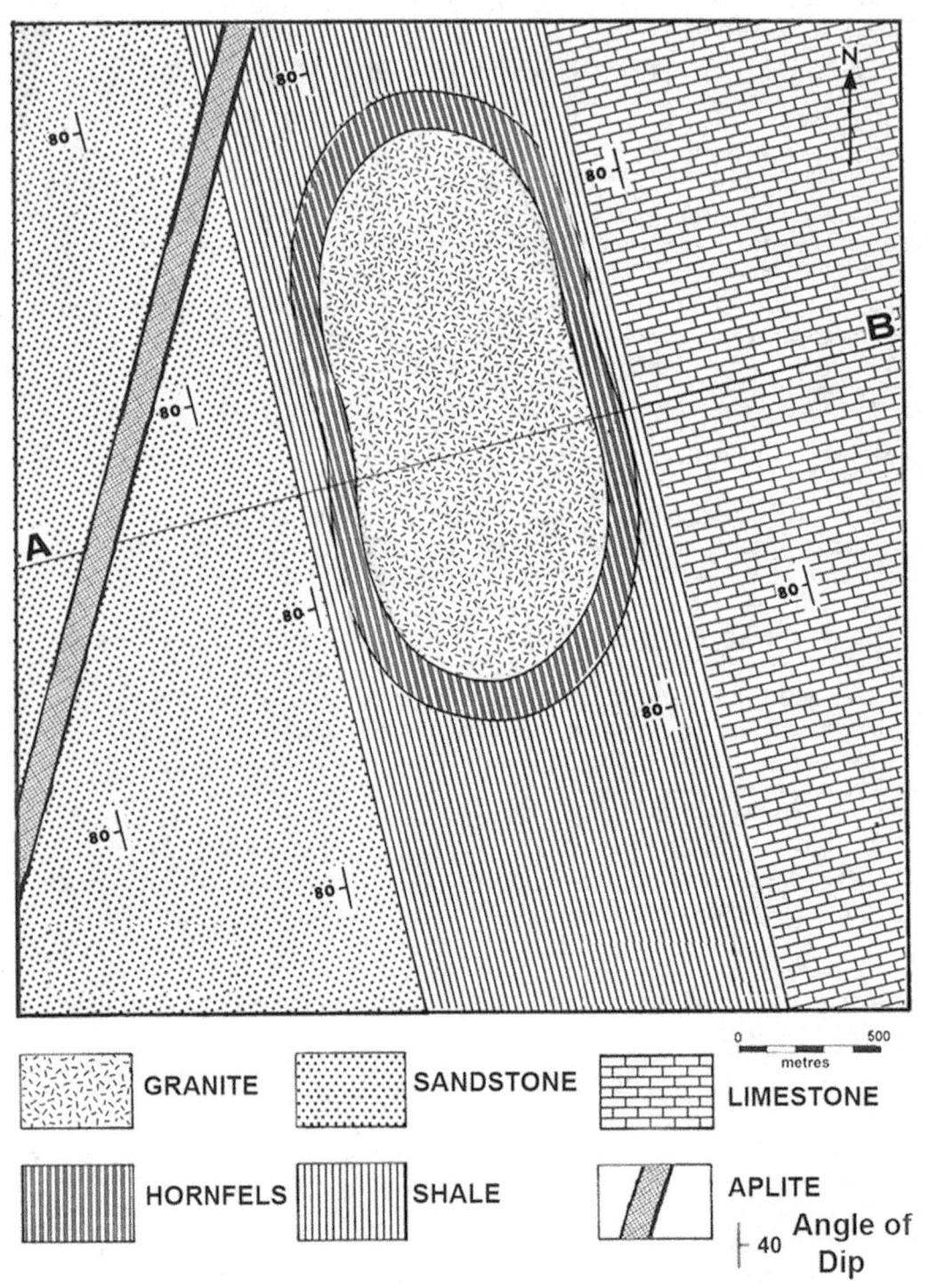

EXPERIMENT 3.5 continued

DATA and OBSERVATIONS:
Draw the cross-section for map showing the intrusions and the surrounding rocks.

CONCLUSIONS:

1. What types of igneous intrusions are seen on this map? Explain the reasons for your answer.

2. What is hornfels? Why does it ring the central body?

3. Why is there no ring or major edge along the dyke?

4. What is the relationship between the rocks granite and aplite?

5. Is there any indication which intrusion came first?

Research (Optional)

1. Use the textbook or the Internet to revise the different types of igneous intrusions.

2. What are some of the commercial uses for some igneous intrusions?

EXPERIMENT 3.6
One lesson

MAPPING UNCONFORMITIES

AIM: To draw a cross-section across a region which contains beds which are separated by a hiatus or break in time (called an unconformity)

MATERIALS: Pencils, rulers, grid/graph paper, paper, protractors

BACKGROUND:
The deposition of sedimentary beds is often not continuous. Sometimes there is a break in deposition caused by such events as lack of sediment or water flow followed by a period of erosion (often with some uplift of the basin floor). If a new cycle of sedimentation occurs in this area, an unconformity is formed.

PROCEDURE:

Look at the plan view (i.e. taken looking down onto the surface) on the next page.

1. Draw a topographical cross-section rectangle where the top is equal to the length A-B and represents the surface. Using an appropriate scale (say 1 cm = 100 metres depth), complete the rectangle so that it represents a depth of 600 metres.

2. Place the edge of a spare paper sheet along the map from A – B and mark off the locations of A, B and the boundaries of each bed. Also mark any angles of dip for each bed and the type of rock represented by the symbols and the position of each of the intrusions where they cut the surface. For simplicity, assume here that these are vertical with dykes being thin with parallel sides and stocks and batholiths coming up from a larger base.

3. Transfer this data to the drawn cross-section rectangle by placing the same paper edge between A and B and marking off the bed boundaries (including their dip) and the position of the edges of the intrusions.

4. Draw lines down through the cross-section for any dykes. If no dip is given for them assume that they are vertical, otherwise draw in the parallel sides of the dyke with the appropriate dip.

5. Draw in dotted lines (of uncertainty) down and spreading out from the surface edges of any stock or batholith to represent that they are a much larger mass below ground. Where these subsurface edges of these intrusions will not be known unless drill cores are taken.

6. Using the protractor along the top of the rectangle (= the land surface), draw in lines at the appropriate dip in each of the different the dip directions to represent any bedding planes which have been tilted and in their dip directions. Lines for tilted sedimentary beds can be carried through to the other sides of the intrusions.

7. Complete the cross-section by adding a small amount of rock symbol (as a representative of the whole bed but care with orientation of the symbol) and the angle of dip for each bed.

EXPERIMENT 3.6

PROCEDURE: continued

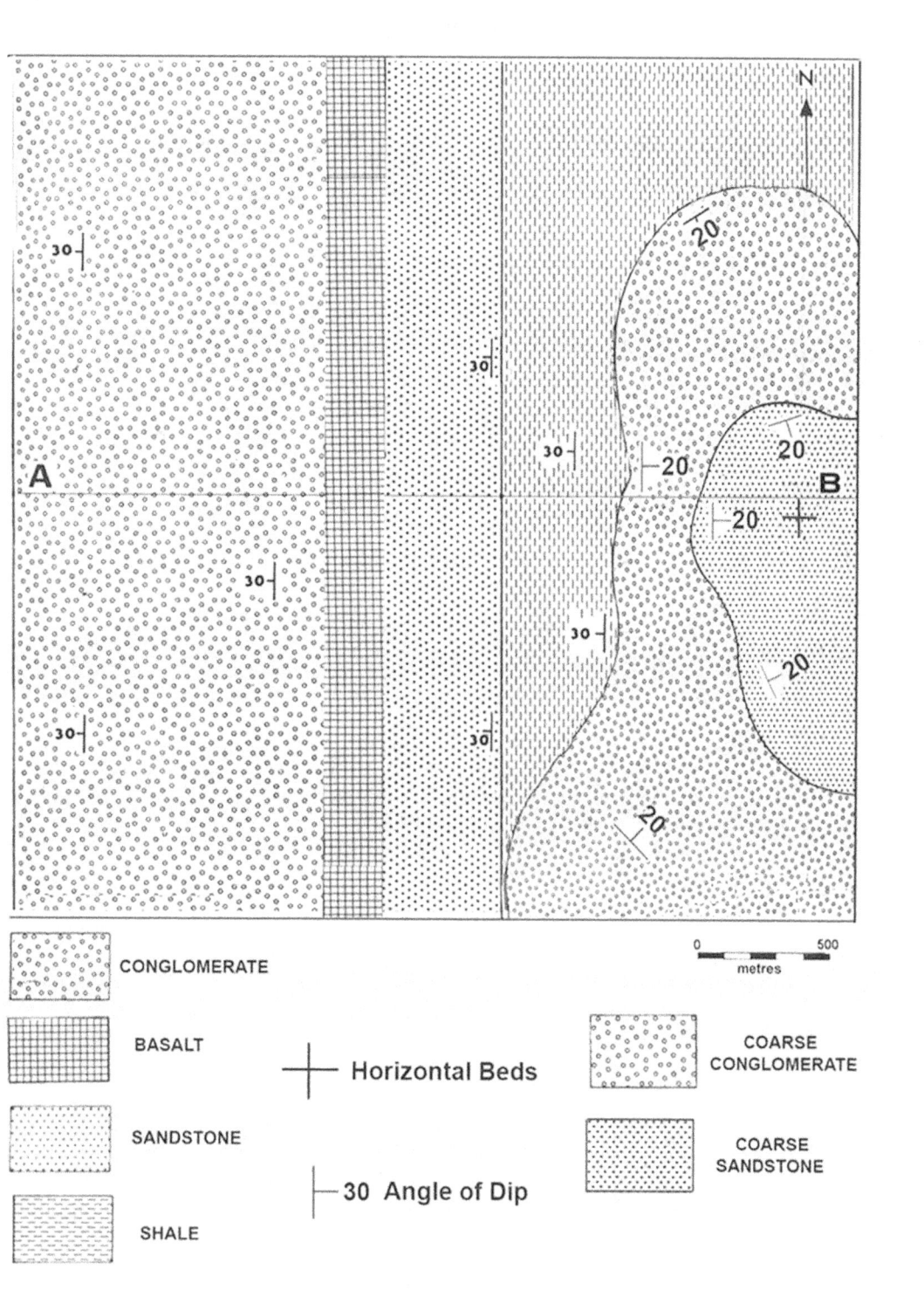

EXPERIMENT 3.6 continued

DATA and OBSERVATIONS:

Draw the cross-section for map showing the unconformities by using a <u>wavy line</u> at the contact of the beds making the unconformity.

CONCLUSIONS:

1. What types of unconformities are shown in the cross-section?

2. What is different about the basalt? How did it get there?

3. Which beds were deposited (a) first and (b) last?

Research (Optional)

Use the textbook or the Internet to revise the different types of unconformities which can occur. Why are they all significant in determining the geological history of an area?

EXPERIMENT 3.7 One lesson

GEOLOGICAL HISTORY

AIM: To draw a cross-section across a region and determine its geological history from the earliest events to the present.

MATERIALS: Pencils, rulers, grid/graph paper, paper, protractors

BACKGROUND:
Geologists use drawn cross-sections and surface geological maps (and much more) to determine what events have occurred in the region from the earliest to current. These will include the deposition of sedimentary beds, intrusions, metamorphic structures formed by Earth movements and erosion.

PROCEDURE:

Look at the plan view (i.e. taken looking down onto the surface) on the next page.

1. Draw a topographical cross-section rectangle where the top is equal to the length A-B and represents the surface. Using an appropriate scale (say 1 cm = 100 metres depth), complete the rectangle so that it represents a depth of 600 metres.

2. Place the edge of a spare paper sheet along the map from A – B and mark off the locations of A, B and the boundaries of each bed. Also mark any angles of dip for each bed and the type of rock represented by the symbols and the position of any fold axes.

3. Transfer this data to the drawn cross-section rectangle by placing the same paper edge between A and B and marking off the bed boundaries (including their dip) and the position of fold axes.

4. Draw construction lines down through the cross-section for any fold axis.

5. Using the protractor along the top of the rectangle (= the land surface), draw in lines at the appropriate dip in each of the different the dip directions to represent any bedding planes which have been tilted and in their dip directions. Lines for tilted sedimentary beds can be carried through to the other sides of the fold axes.

6. Complete the cross-section by adding a small amount of rock symbol (as a representative of the whole bed but care with orientation of the symbol) and the angle of dip for each bed.

7. Review the entire completed cross-section and devise a geological history for all beds and structural events. Use all of the information learned about how specific rocks are formed and how geological events happen to include in this detailed history.

EXPERIMENT 3.7

PROCEDURE: continued

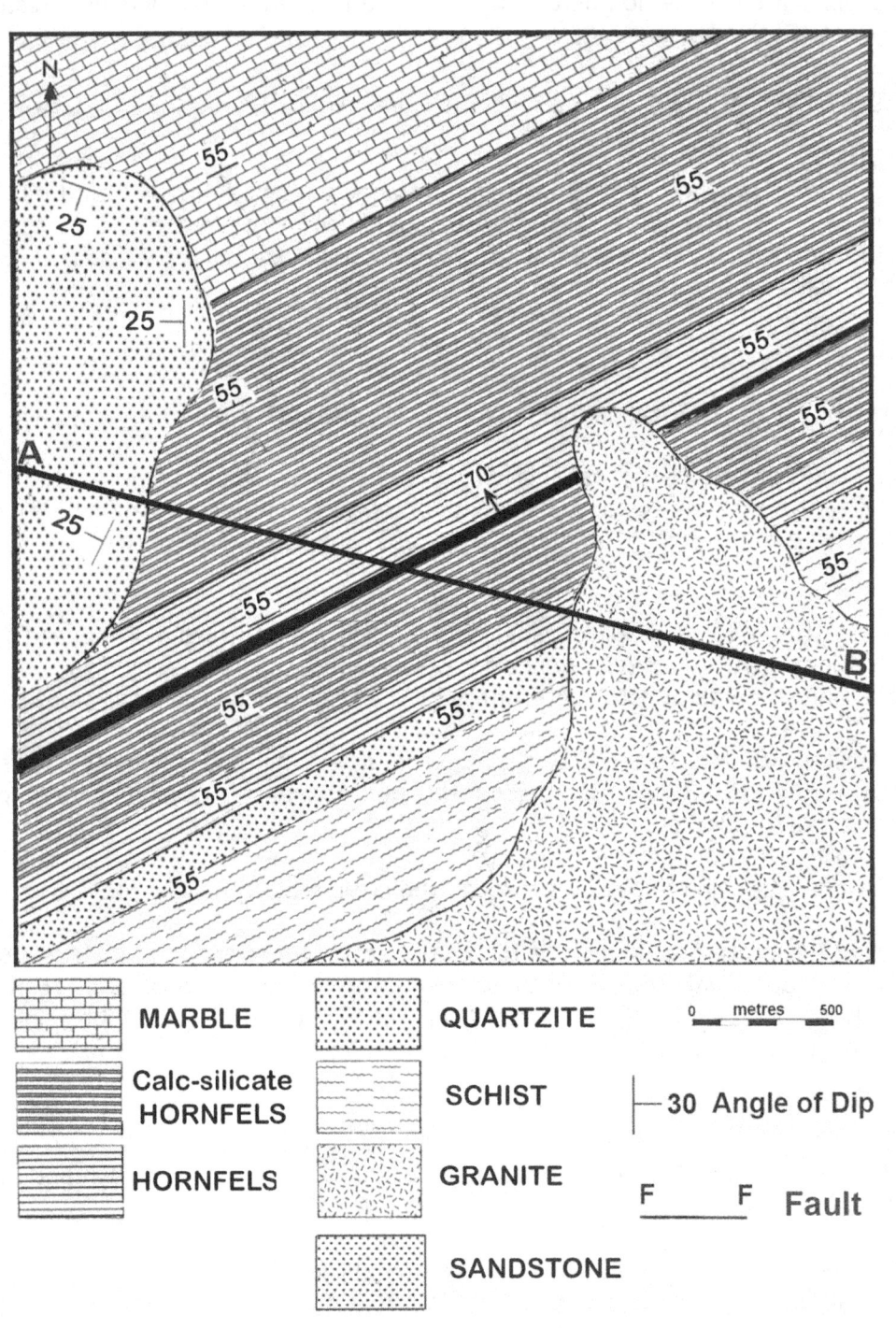

EXPERIMENT 3.7 continued

DATA and OBSERVATIONS:

Draw the cross-section for map showing, faults, unconformities (use a <u>wavy line</u>) and intrusions.

CONCLUSIONS:

1. What types of unconformity and fault are shown in the cross-section?

2. What is the (a) heave and (b) throw of the fault?

3. What is different about the layered beds? How were they originally formed?

4. List in order all of the events which occurred in this region from the first event to the current day (care: this is a complex problem so requires some research and in-depth description).

Research (Optional):

Use the Internet or local resources (maps, books, experts etc.) to find out about the local geology.

EXPERIMENT 3.8 One lesson

HALF-LIFE OF RADIOACTIVE IODINE 131

AIM: To use experimental data to determine the half-life of radioactive iodine.

MATERIALS: Data (supplied), pencils, rulers, grid/graph paper.

BACKGROUND:
Radioactive isotopes decay by emitting radiation as particles and electromagnetic gamma rays producing daughter products and decreasing their own mass. Half-life is the time taken for a certain mass of an isotope to decay to only half of its original mass. There would also be masses of various daughter products as well.

To measure half-life, a known amount of radioactive isotope is placed near Geiger-Muller Counter which detects the radiation coming from the specimen as a series of audible clicks through a loudspeaker. An automatic counter also shows the number of counts per second.

PROCEDURE:

1. Examine this table of data taken from a typical measurement of iodine -131 over a number of weeks:

DAY	COUNTS PER SECOND
0 (start)	10,000
7	5,500
14	3000
21	1800
28	900

2. Use the graph paper (or a graph App) to draw the graph of this decay. Use Counts per second on the vertical axis and the number of days on the horizontal axis.

3. Join the plotted points by a curve of best fit.

4. Find the half-life by estimating that place on the curve where the number of counts per second has gone from the maximum amount to exactly half of this amount and hence find the corresponding time for this drop in counts per second.

EXPERIMENT 3.8 continued

DATA and OBSERVATIONS:

Draw the half-life graph here and show all working used in determining the half-life of Iodine-131

CONCLUSIONS:

1. What is the graphical value of the half-life for iodine-131?

2. What could be some of the errors involved in this experiment?

3. If 40 grams of iodine-131 was used at the start of the experiment, how long would it take for this mass to decay to 5 grams of iodine-131?

Research (Optional):

Use the Internet or local resources (maps, books, experts etc.) to find out about the uses of radioactive iodine-131.

EXPERIMENT 3.9   One lesson

SIZE OF THE EARTH

AIM: To find the circumference of the Earth.

MATERIALS: metre rule, shadow stick, Internet and computer or tablet

BACKGROUND:

This method and its geometry were well-known to the Ancient Greeks, notably Eratosthenes of Cyrene (276-195 BC). Cyrene is now Aswan in southern Egypt and Eratosthenes was the Chief librarian at the great library at Alexandria in Egypt (then a Greek colony). He knew that at local noon on the summer solstice in Cyrene, the Sun was directly overhead because the shadow of someone looking down a deep well at that time blocked the reflection of the Sun on the water. He measured the Sun's angle of elevation at noon in Alexandria by using a vertical rod (a shadow stick or gnomon) to measure the length of its shadow. Knowing the length of the rod, and the length of the shadow, he calculated the angle of the sun's rays at Alexandria to be about 7° (1/50th of 360° and so proportional to that part of the circumference of a circle). Assuming that the Earth as spherical (having seen its circular shadow on the Moon during a lunar eclipse), and knowing both the exact distance across the ground to Cyrene (a well-known trade route), he concluded that the Earth's circumference was fifty times that distance.

PROCEDURE: (WARNING: Never look directly at the Sun)

1. Place a shadow stick of known height in the centre of an open area exposed to direct sunlight ensuring that the stick is vertical.

2. Measure the length of the shadow along the ground.

3. From the length of the shadow stick and the length of the shadow, calculate the angle of the Sun by using trigonometry by:

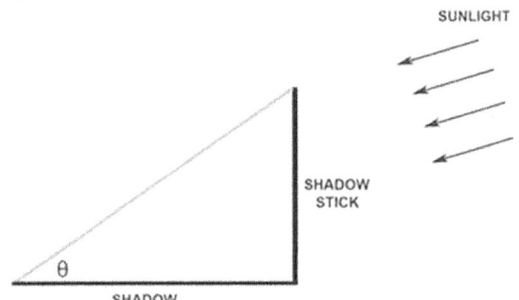

   (a) finding the tangent of this angle (θ) by dividing the set height of the shadow stick by the length of the shadow.

   (b) finding the inverse ($\tan^{-1}$) of this value from tables or an Internet converter. (e.g. at https://www.rapidtables.com/calc/math/Tan_Calculator.html)

EXPERIMENT 3.9 continued

PROCEDURE: continued

4. Use an atlas or Internet maps to find another location north (or south in the northern hemisphere) which is on or near the equator but with approximately the same longitude (i.e. is exactly due north).

5. Use a website to find this new locations Sun angle as approximately equal to the same time as the measured angle taken here.

    https://www.timeanddate.com/sun/australia/brisbane?month=3
    (type in the location of the place on the equator)

6. Measure the distance (in kilometres) between the observers location and this new location on the equator using a map or typing "distance between [x] and [y]" into a search engine where x and y are the two places of interest.

7. Find the difference in the angles of the Sun between these two positions. This corresponds to the angular difference between the two places at this known distance.

8. Use ratios to calculate the circumference of the Earth if this angular distance was 360^0

DATA and OBSERVATIONS:

Record each of the following:

> Height of the shadow stick (cm.)
>
> Length of the shadow (cm.)
>
> This location (latitude and longitude)
>
> Calculated Sun angle (degrees)
>
> Time of day
>
> Location of place on the Equator (give latitude and longitude)
>
> Angle of the Sun at this place on the Equator

CALCULATIONS:

Now the ratio of the angle difference ($\Delta\theta$) to 360^0 should be the same as the ratio of the land distance between the two places compared to the circumference of the Earth (C).

$$\text{i.e.} \quad \Delta\theta / 360^0 \quad = \quad D/C$$

$$\text{or the circumference} \quad C \quad = \quad 360 D / (\Delta\theta)$$

EXPERIMENT 3.9 continued

CALCULATIONS: continued

Where:
(Δθ) = Difference in Sun angles between the two places
D = Distance between these two places (kilometres) and
C = Circumference of the Earth (using North South radius)

The real value for the circumference around the poles is 39, 931 kilometres.

CONCLUSIONS:

1. What is the calculated value for the circumference of the Earth (along north south)?

2. How does this compare with the actual value of 39,931 kilometres? Calculate the percentage error for this experiment.

3. How could this experiment be improved?

Research (Optional): Use the Internet to find out:

1. How the actual size and shape of the Earth (geodesy) has been determined with modern equipment.

2. More about Eratosthenes of Cyrene (Syrene)

Chapter 4: Rock-forming Minerals

EXPERIMENT 4.1   One to three Lessons

SOME COMMON ROCK-FORMING MINERALS

AIM: To describe, draw and then identify some common minerals from which rocks form.

MATERIALS: Mineral specimens (quartz; orthoclase; plagioclase; calcite; hornblende; olivine in basalt rock; biotite; muscovite) hand lens, streak plates, and Mohs' scale of hardness kit.

BACKGROUND:

Minerals are pure, inorganic chemical compounds having definite properties. These properties which are useful in identifying specimens include:-

1. COLOUR - the frequencies of light coming from the specimen when viewed in normal white light. Simple colours (e g black, red-brown etc.) are used to describe this property. A mineral may have a variety of colours.

2. STREAK - the colour of the powdered mineral. If soft, the specimen may be quickly drawn across a streak plate (an unglazed tile -white for coloured minerals, black for white or clear minerals). If the specimen is hard, it may be scratched with a knife or a harder mineral and the colour of the scratch observed.

3. CRYSTAL FORM (Habit) – the overall appearance of the specimen in the way that it has been formed e.g. cleavage block (individual crystals not seen, but many flat crystal surfaces observed giving a blocky appearance); botryoidal (many rounded lumps together); fibrous (long strands or fibres); radiating (long crystals radiating outwards from a common point); granular (grouped like grains of sand); amygdaloidal (crystals growing in holes in rock); foliated (as sheets); pisolitic (pea-sized spheres); or massive (no crystals seen).

4. LUSTRE - the way light reflects off the specimen's surface. It may vary from surface to surface (i.e. a specimen may have a range of lustre). Lustre may be:

        Metallic -hard, shiny like a metal
        Shiny – very reflective like gloss paint
        Vitreous -glassy, with some depth
        Pearly -like a pearl or button
        Silky -shiny with a fibrous look
        Greasy -oily appearance like soap
        Resinous -with a dull transparent look
        Dull -very little shine

EXPERIMENT 4.1 Continued

5. CLEAVAGE - the ability to break naturally along flat surfaces. The best test is to actually cleave the mineral but this is often not desirable so one has to look for the number of cleavage planes which may meet together. Minerals may have:

>No cleavage - but the mineral may have crystal faces e.g. Quartz or cleavage in:

>One direction called basal cleavage which gives sheets
>Two directions giving lines of small steps
>Three directions giving points or corners (which also may be cut off) or even in
>Four directions giving four-cornered points like a pyramid (rare).

6. HARDNESS - resistance of the mineral to be scratched. Usually compared to a standard set of minerals (Mohs' Scale) or a set of material for approximations (Field Scale):

Mohs' Scale	Field Scale
1. Talc (Softest)	Thumb nail 2.5
2. Gypsum	Coin 3.5
3. Calcite	Glass 5.0 -6.0
4. Fluorite	Knife 5.5 -6.0
5. Apatite	File 6.5- 7.0
6. Orthoclase	
7. Quartz	
8. Topaz	
9. Corundum (Ruby)	
10. Diamond	

7. SPECIFIC GRAVITY - the density of the mineral compared to that of water i.e. how many times the mineral in heavier than an equal volume of water (density of water = 1g/cc). Heft is a word to generally describe the mineral's heaviness in vague terms e.g. heavy, medium, light).

8. OTHER PROPERTIES - any other unique observation e.g. flexible, fluorescent, radioactive, magnetic, chemical reaction, taste (careful!) etc.

9. SKETCH:

A sketch is to be drawn for selected minerals which clearly show some of the above features. These sketches should be to scale and in appropriate colour. This is done by:

a. Selecting an appropriate orientation (side) from which the sketch is made.

b. Studying the outline and then drawing it lightly to scale so that about two or three sketches will fit per page (you may put a border around each if you wish).

EXPERIMENT 4.1 Continued

    c. Re-examination of the specimen (possibly with a hand lens) to determine the main internal features for exaggeration and drawing these within the outline. If there is considerable detail, only draw a representative part of it.

    d. Once the sketch is complete, go over the outline and main features with heavier pencil (or even black ink if you are careful).

PROCEDURE:

1. Draw up a **table** (with a full page turned sideways) under data & observations using the following headings:
      code   colour   streak   lustre   cleavage   hardness   other

2. Test and describe each specimen for each of the properties (given above).

3. Draw each specimen to scale (e.g. x2 or x3 etc.) and **in colour** where appropriate.

4. Use the text, display minerals, computer and the key (provided below) to name the specimen. Warning: not all the minerals may be in the list below!

DATA and OBSERVATIONS: (Table and Sketches as appropriate)

CONCLUSIONS:

1. List the code numbers of the specimen, the specimen's names and at least two distinctive properties of each mineral which will then help in the quick identification of that mineral.

2. Comment on the use of mineral properties (advantages/disadvantages) in identifying unknown minerals.

3. What are some likely errors which may limit correct identification of minerals in the field?

EXPERIMENT 4.2   One lesson to set up then several days

GROWING CRYSTALS

<u>AIM:</u> To grow well-formed crystals from solution.

<u>MATERIALS:</u> Copper sulfate or alum or chrome alum; petri dishes; de-ionized water; salicylic acid crystals; beakers (250 ml); sticks; thread; large test-tube & rack; Bunsen burner or hotplates.

<u>BACKGROUND:</u> Crystals can form from evaporation of water from a mineral solution or by cooling a hot solution down until the solute "un-dissolves" and precipitates as crystals. Once a small crystal has formed and is still in a saturated solution, it may continue to grow as more ions add to it from the solution (accretion).

<u>PROCEDURE:</u>

PART A: CRYSTALS FROM HOT SOLUTIONS

Care! Heating water in a test-tube can be dangerous if done too quickly. Use paper to hold the tube, point it away from others and heat slowly on a low flame.

1. Add a small amount (about half a spoonful) of salicylic acid (a solid organic acid used in making aspirin) to about half a **large** test-tube.

2. Carefully boil the water and dissolve the solid.

3. When completely dissolved, hold the outside of the test-tube under cold water and observe carefully.

<u>QUESTIONS:</u>

1. What happens as the solution cools? Why?

2. Describe any shapes. Sketch the apparatus and observations in the Results section.

PART B: GROWING LARGER CRYSTALS

1. Dissolve copper sulfate crystals (or alum or chrome alum) in a little hot water - enough to fill a small petri dish.

2. Pour the solution in the dish and leave overnight in a forced-air fume hood or place of good air circulation.

3. Examine the shapes and any other feature using a hand lens or microscope.

4. Select the <u>best shaped</u> crystal and secure it with a loop of cotton thread, so that when attached to the stick it can hang in the beaker about 1-2 cm. from the bottom (see diagram next page).

EXPERIMENT 2.2 Continued

5. Cover the beaker with paper (why?) and leave overnight.

6. Each day thereafter (for about a week), top up the solution (it must always be saturated. Why?) and check the size and shape of the crystal. Remove any imperfections or smaller crystals on the string.

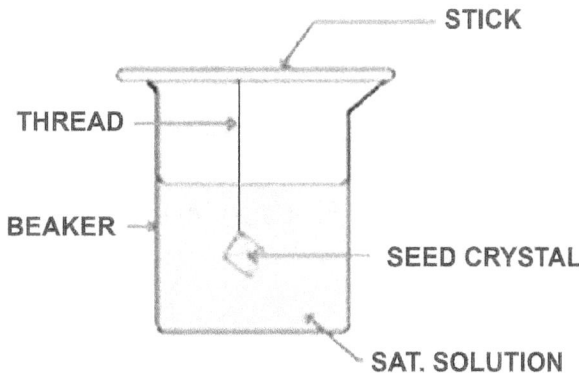

7. Record the size (e.g. width) of the crystal each day (in millimetres) and enter the data in a table in results.

DATA and OBSERVATIONS:

Draw the larger crystal which has been grown with an appropriate description

CONCLUSIONS:

1. Comment generally on how these crystals can grow from solution and how smaller crystals grow to bigger size.

2. What are the limitations to the size of the crystals?

3. Is there any indication within the crystal as to how they grow from smaller crystals?

4. Where in nature do crystals grow in this way (Internet research needed).

EXPERIMENT 4.3   One lesson

MEASURING SPECIFIC GRAVITY

AIM: To determine the specific gravity of a mineral.

MATERIALS: mineral specimen; electronic balance; large beaker (250 ml); thread

BACKGROUND: Specific gravity is the density of the mineral compared to the density of water (= 1 gram/cubic centimetre   Note: 1 cc = 1 millilitre approx.)

$$\text{i.e. SG} = \frac{\text{Density of Mineral}}{\text{Density of Water}}$$

(note: there is no unit of measurement – it is a ratio)

$$\text{density} = \text{mass (in grams)}/\text{volume (in ml)}$$

PROCEDURE:

1. Weigh each specimen carefully in air on the balance and the note the value.

2. Half-fill a 250 ml beaker with water and place it on the balance which was been zeroed (tared).

3. Secure the specimen by a loop of thread and completely submerge it below the water in the beaker. The reading on the balance represents the weight of water displaced by the specimen and thus its **volume** (Archimedes' Principle).

4. Repeat the measurements several times and take an average.

5. Using the mass and this value for volume, calculate the density of the mineral. Since S.G. is a comparison to the density of water which is 1 g/cc then this numerical value calculated is the S.G. of the mineral.

6. Using the average value of the S.G. and the table (given at the end), identify the white mineral

7. Record all data and any observations Also give the error of any measurement

Remember that for any analogue instrument (where all divisions can be seen), error = half a unit for any digital instrument (only one unit can be seen) = one unit

When calculations are made using errors, it is best to use percentage errors which are then added to give the overall error for the calculated answer.

e.g. % error for mass   = $\dfrac{\text{instrument error}}{\text{measured value}} \times 100$

To be more accurate, the measurements can be taken several times and calculate the S.G.

EXPERIMENT 4.3 Continued

CALCULATIONS: Use the formula given in Background to calculate S.G. and give an overall % error

CONCLUSIONS:

1. What was the calculated value for S.G. and its error?

2. From a list of S.G. what is this mineral?

3. How does the experimental value agree with the book value for this mineral?

4. Comment on any student, environmental or instrument errors.

5. How could this experiment be improved?

TABLE OF SOME SPECIFIC GRAVITIES

Mineral	S.G.
Borax	1.7
Halite	2.1
Calcite	2.7
Plagioclase	2.6
Aragonite	2.9
Strontianite	3.8
Barite	4.5
Anglesite	6.3
Cassiterite	7.0

EXPERIMENT 4.4   Two lessons

GEOCHEMISTRY

AIM: To observe and describe some common tests for metal ions and some salts and hence determine the chemistry of a given unknown mineral.

MATERIALS: Bunsen burner & ceramic mat; paper clips; wooden tongs; test-tubes & racks; beaker; limewater solution; safety equipment (apron, glasses, gloves); powdered metal salts (e.g. copper sulfate, sodium chloride, barium carbonate, strontium nitrate, lithium chloride); powdered minerals e.g. sulfides (galena); carbonates (dolomite, calcite, magnesite); chloride (halite); and sulfate (epsomite); 2M Hydrochloric acid in dropper bottle; dropper bottles of barium nitrate solution and silver nitrate solution.

BACKGROUND:

Chemical testing of various minerals has been on-going for a long time. At the end of the 19th Century, German chemists (notably Bunsen and Kirchhoff) were using the newly discovered spectroscope to analyse the light coming from coloured flames produced by burning metal salts in a gas flame. This later led to the development of the Flame Spectrometer which is used today to identify the common metals in minerals. Sometimes the mineral specimens are too small to identify by physical properties, so chemical testing is required.

Several ion groups can also be tested in the school laboratory. Carbonates give odourless carbon dioxide gas in acid and sulfides give smelly hydrogen sulfide (very poisonous! Do not smell too much of it!) with acid. Chlorides give a white precipitate (of silver chloride) with silver nitrate solution (stains the skin!) and <u>soluble</u> sulphates give a white precipitate (of barium sulfate) with barium nitrate.

PROCEDURE:

PART A: FLAME TESTS (one lesson)

1. Set up a Bunsen burner so that it sits on a protective ceramic mat.

2. Unfold a paper clip to a straight wire and, holding one end in wooden tongs, heat the other end in the <u>blue</u> flame of the Bunsen burner.

3. Dip the hot end of the wire into water in a beaker and then dip it into a sample of a metal salt to pick up one or two small crystals (important use the smallest possible!).

4. Heat the crystal(s) in the <u>blue</u> flame of the Bunsen burner and note the colour of the flame.

5. Repeat parts 3 and 4 (above) for each of the selection of metal salts. Remember to <u>clean</u> the end of the wire in water after each test and note the colour of the flames for the metal salts (copper sulfate, sodium chloride, barium carbonate, strontium nitrate, lithium chloride).

EXPERIMENT 4.4 continued

PART B. CHEMICAL TESTS (Lesson 2)

TEST FOR CARBONATES

1. Into separate test tubes place a small amount (half-pea size) of each of the metal carbonates provided (dolomite, calcite, magnesite). Remember the positions and names of each sample.

2. Just cover each sample with a little acid (CARE! It is caustic and will damage tissue) and observe any reaction (cautiously sniff any gas, waft a little towards the nose using your hand in a wave like action....there may also be some pungent acid vapour).

3. Place a glass rod which has been dipped into some limewater and note any changes.

TEST FOR CHLORIDES

1. Into a test tube place a very small amount of Halite (Sodium Chloride).

2. Cover with about half a centimetre of water and shake to dissolve some of the mineral.

3. Add two drops of silver nitrate solution (CARE!) and observe.

4. Note any reaction.

TEST FOR SULFATES

1. Into a test tube place a very small amount (2-3 rice grains) of epsomite (magnesium sulfate);

2. Cover with about half a centimetre of water and shake to dissolve some of the mineral;

3. Add two drops of silver barium nitrate solution and observe;

4. Note any reaction.

TEST FOR SULFIDES (CAUTION! - poisonous gases)

(Students are warned not to make this gas in large amounts and to sniff it very cautiously. IMMEDIATELY after all students in the group have made an observation, the test tube and its contents must be handed in for removal to the fume hood)

1. Into separate test tubes place a very small amount (about the size of a rice grain) of the lead sulfide (e.g. galena).

2. Cover with about half a centimetre of acid and observe. CAUTIOUSLY smell any gas given off by waving a hand across the top of the test-tube towards the nose.

3. Hand in the test-tubes for placement into a fume hood (or outdoors) for disposal in a bucket filled with water.

EXPERIMENT 4.4 continued

RESULTS:

1. Record all of your observations in a table;

2. Find out the chemical name of each mineral tested and write a word equation for each reaction;

3. draw a representative sketch (one half to one page in two dimensions with labels) of any one of the test-tube reactions.

CONCLUSIONS:

Use the following questions to write a detailed conclusion about this activity:-

1. What was the test and it's result for each metal ion (flame test) and for each non-metal group (carbonates etc.)? list or table summary;

2. Why was de-ionized water used for making up solutions and dissolving the minerals?

3. What would be the main errors in these tests (be specific for parts a & b)?

4. in what situations may a geologist in the field use some of the tests in part b? explain.

RESEARCH: (Optional)

What are some of the instruments and methods used by Geochemists to analyse specimens to find their composition.

Chapter 5: Igneous Rocks - The Beginning

EXPERIMENT 5.1   One or two lessons

SOME COMMON IGNEOUS ROCKS

AIM: To examine and describe some common igneous rocks.

MATERIALS: Hand lens, rock specimens of granite, gabbro, andesite, trachyte, diorite, rhyolite, basalt, a porphyry, pumice, tuff and a volcanic glass.

PROCEDURE:

1. Examine each specimen in turn using the hand lens.

2. Describe each rock in turn using the following descriptors and a table to record the data in Results:

COLOUR: overall colour appearance e.g. pink with black spots, grey, black etc.

CRYSTAL SIZE: as coarse (> 2mm), fine (<2 mm), glassy (none) or porphyritic (has bigger crystals in a background of smaller crystals).

TEXTURE: is the way the crystals appear to be locked together e.g.

> phaneritic – all crystals large (> 2mm) and interlocked
> aphanitic – crystals mostly not visible to the naked eye but can be seen under a microscope.
> porphyritic – big crystals (called phenocrysts) in a smaller crystalline background (or matrix). indicates two stages of cooling.
> glassy (holohyaline) no crystals seen, not even under a microscope.
> pyroclastic broken and angular fragments formed by volcanic explosion.

COMPOSITION: if you can see crystals, NAME them using simple common terms:

> e.g. quartz looks glassy grey
> orthoclase is pink and blocky
> plagioclase is white or shiny blue grey – may be long and shiny in dark rocks
> biotite is black and very shiny flakes
> olivine is dark green and glassy granules like sugar
> hornblende is black, thin and long.

> If no crystals seen (often = very fast cooling) then write "none seen"

> **STRUCTURES:** what shapes can be seen within the rock (mostly in Extrusive rocks) e.g.
> massive – no structure, only uniform crystals or no crystals at all
> vesicular – gas bubbles
> amygdaloidal – gas bubbles filled with mineral.
> fluidal flow lines (may look like part layers)

EXPERIMENT 5.1 Continued

SKETCH - draw three sketches of the specimen which show:

    Phaneritic texture with large crystals, mostly equal sizes. Name and label the main minerals seen in this specimen;

    Porphyritic texture of large crystals within a background of smaller crystals. Name and label the larger crystals (phenocrysts); and

    Holohyaline texture with no crystals but showing structures such as flow lines.

DATA and OBSERVATIONS: Construct a table of data from descriptions and sketches to scale in colour. Sketches should be of a suitable size e.g. two or three to a page.

Code	Colour	Texture	Mineralogy	Grain Size	Structure
				(each mineral)	(if seen)

CONCLUSIONS:

1. In your Conclusion, try to name each of each specimen and for each, give a few key words to remember each rock e.g. pumice could be grey, bubbles.

2. Comment on the significance of each rock i.e. how (and where) it might have been formed. That is, classify each rock as either:

   - intrusive (formed below surface) or extrusive (formed on surface) and
   - felsic light in colour) or mafic (dark in colour) e.g. rock No. # is intrusive, mafic etc.

RESEARCH: optional

1. Find out about how igneous rocks are used. Try to be as detailed as possible.

2. Why are do some igneous rocks produce good soils? Often these are on the slopes of very active volcanoes yet people still live and farm there. Make a list (or case study for group display) of some of the world's most active volcanoes which are populated and have caused considerable devastation to these populations. Have they been re-populated?

EXPERIMENT 5.2

One or two lessons

IGNEOUS ROCKS TEXTURES and COMPOSITION EXERCISE

AIM: To examine some photomicrographs and (a) determine the textures of those thin-sections shown and (b) do a point count on one sample and estimate its mineral composition.

MATERIALS: Photomicrographs of various igneous rocks.

BACKGROUND:

Thin-sections of rocks are made by slicing them with a diamond saw to make slabs about a few millimetres thick. These are then turned upside down and slowly ground down using different abrasive powders and water until they are about 30 microns (µm) thick. This makes the rock section transparent. Placed below a petrological microscope between polarising filters, the texture i.e. how the crystals fit together, is easily seen and the minerals have characteristic structures, shapes, degrees of internal clarity and colour. Even in black and white photos of these thin sections (called photomicrographs) minerals can still be identified.

Using a grid across the slide or a device which moves the slide across and down at exact intervals (like a grid), an estimate of composition can be found by tallying the mineral which falls under a crossed part of the grid. This is a form of point count.

PROCEDURE:

PART A: TEXTURES

1. Examine the photomicrographs below and use the notes and diagrams from the textbook to identify their igneous rock textures.

2. List each letter of the specimen and describe its texture in detail. Also suggest a mode or place of formation of this rock and record the details in a table showing the identifying letter, the texture and the mode of occurrence and formation and an hypothesis of the name of the specimen.

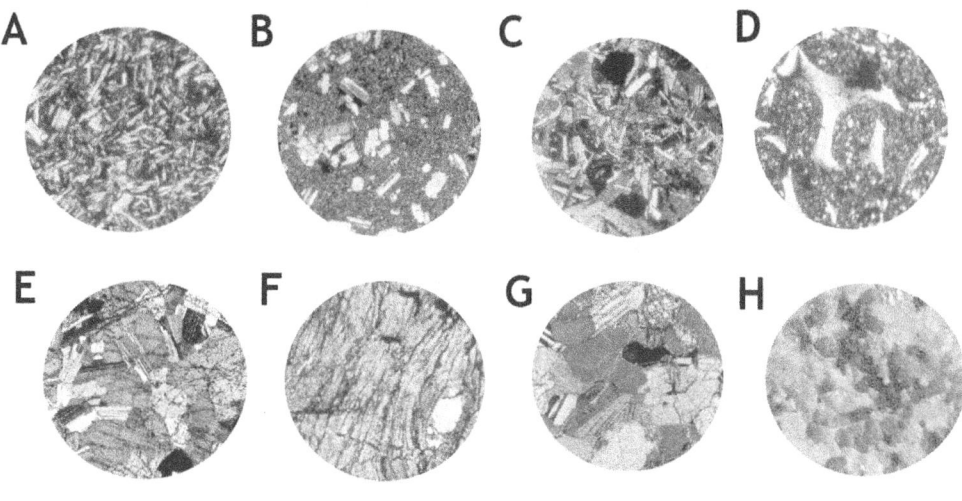

MAGNIFICATION X 25 CROSS POLARS

EXPERIMENT 5.2 Continued

PART B: Composition Estimation using a Point Count

1. Examine the photomicrograph of an igneous rock which has been place under a grid.

2. Identify the minerals under each set of crossed lines and hence estimate the percentage composition of the rock. This is done by placing the tally for each mineral over the total number of tallies then multiplying by 100/1 to get a percentage.

3. Refer to the table in the text book showing percentage composition of the main minerals and attempt to name the rock.

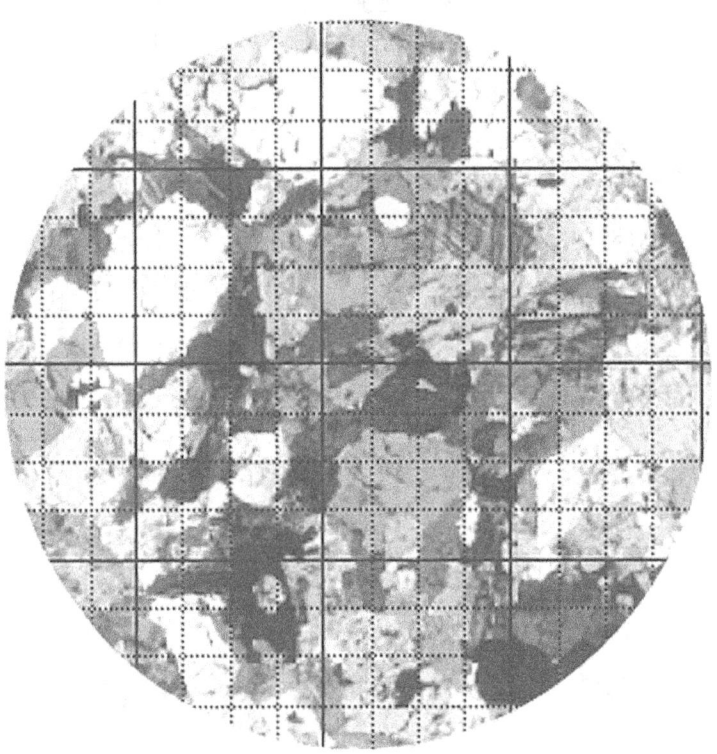

EXPERIMENT 5.2 Continued

Identification Chart of Some Common Minerals

AUGITE 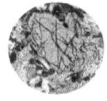 Has irreglar crystals often cloudy with many lines and dots. Darker shades.

BIOTITE MICA 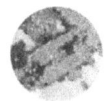 Long sheets of darker and cloudy shade often with wavy parallel lines.

HORNBLENDE  Irregular and well-defined crystals, relatively light and clear shades.

MUSCOVITE MICA  Long sheets similar to Biotite but lighter in colour.

OLIVINE  Ill-defined crystals often clouded and with many criss-crossed lines.

OPAQUES e.g. Magnetite 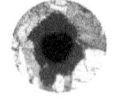 Uniformly black. Magnetite may be seen as small squares due to cubic cleavage.

ORTHOCLASE  Blocky crystals, relatively clear and light in shade but often has dark inclusions.

PLAGIOCLASE  Blocky but usually light and clear. Often shows twinning as two or more sections of different shades together

QUARTZ  Rounded or blocky and almost always light and clear.

DATA and OBSERVATIONS:

PART A: Construct a table of showing the identification letters, textures and if possible the names of each rock.

PART B: Draw up a table showing a rough tally for each of the main minerals seen (warning: not all minerals in the chart may be seen!) and give the percentage composition for each mineral. Refer to the text book chapter on Igneous Rocks to identify the name of this rock (hint: look at its texture).

EXPERIMENT 5.2 Continued

CONCLUSIONS:

1. What were the textures and names for each of the rocks in Part A?

2. List the percentage composition of the minerals in the rock of Part B and name the rock.

3. The error in this experiment is the poor quality of the photomicrographs, however under ideal conditions, better views at x 25 magnification and rotatable cross-polars also gives distinguishing colours and colour changes for each mineral. Comment on how this technique would be useful in the identification of ALL rock types.

RESEARCH: optional

Use the internet to locate sites showing good photomicrographs of the names given to the rocks in this experiment and make comparisons.

Chapter 6: Sedimentary Rocks

EXPERIMENT 6.1:   One Lesson

SOME COMMON SEDIMENTARY ROCKS

<u>AIM:</u> To examine and describe some common sedimentary rocks.

<u>MATERIALS:</u> Bottles of sediment and water, collection of sedimentary rocks (e.g. conglomerate, sandstone, shale, mudstone, greywacke, coal and limestone), hand lenses, watch or clock with second hand. Small dropper bottles of dilute acid.

<u>PROCEDURE:</u>

PART A: FORMATION OF LAYERS FROM SEDIMENTATION

    1. Closely examine the bottles of sediment and water provided (DO NOT remove the tops). Ensuring that the top is secure, shake the bottle firmly to mix all of the sediment. What does this represent in nature?

    2. Stop shaking the bottle and stand it upright on the table. As soon as you stop shaking it, time how long for each type of sediment to stop falling. The biggest sediment is gravel, the next is sand, the third is silt and the rest (which may stay **suspended** in the water for a long time) is clay (this might take several **days** to settle).

    3. Record these times in a table in the Results (Part A).

How have the sediments settled (in what order)? Record the answer in the Results and draw a sketch of the sediments in the bottle (a two-dimensional side view).

PART B: DESCRIPTION OF SPECIMENS OF SEDIMENTARY ROCKS

    4. Closely examine each rock specimen and describe your observations in a table in Results (Part B). Use the following descriptors:

        COLOUR general overall colour
        GRAIN SIZE as **coarse** (>2mm), **medium** (1mm - 2mm) **fine** (<1mm)
            or **none** (see the table of the Wentworth Scale provided)

Wentworth Scale

GROUND	SIZES	TYPICAL ROCK
Boulders	Greater than 256 mm	Conglomerate (rounded) or Sed. Breccia (angular)
Cobbles	64 - 256 mm	
Pebbles	4 - 64 mm	
Granules	2 - 4 mm	
Very coarse sand	1 - 2 mm	Sandstone
Coarse sand	0.5 - 1mm	
Medium Sand	0.25 - 0.5 mm	
Fine Sand	0.125 - 0.25 mm	
Very Fine Sand	0.062 - 0.125 mm	
Silt	Less than 0.062 mm Looks smooth Feels gritty	Siltstone

EXPERIMENT 6.1 continued

GRAIN SHAPES will often indicate the amount of transportation which the grains have undertaken as well as their hardness. Use the following diagram:

SHAPE - will vary from VERY ANGULAR to WELL-ROUNDED

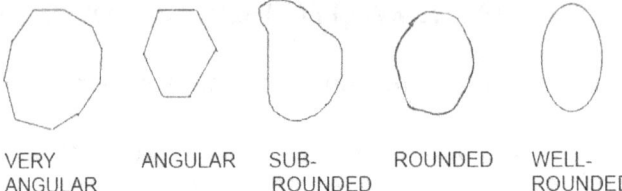

VERY ANGULAR   ANGULAR   SUB-ROUNDED   ROUNDED   WELL-ROUNDED

SORTING shows the arrangement of the grains as:

    well-sorted – all grains about the same size
    medium sorted – most grains about the same size
    poorly sorted – most grains of different sizes;

COMPOSITION - the nature of the rock's grains (clasts or particles), matrix (particles between the grains) and cement. Try to estimate the relative percentage of the composition of the grains. Grains (and matrix) are usually quartz, feldspar or rock fragments (rock fragments give lithic sandstones, conglomerates etc.). Cements are usually calcite (bubbles with acid), silica (i.e. quartz which gives a hard rock) or iron oxides (gives a brown or yellow colour).

STRUCTURES any layers, fossils, bigger grains etc. The most common structures seen are:
    lamina or fine layers
    graded bedding – big clasts ranging (upwards) to smaller clasts but usually only seen in larger scale
    massive – no structures, all uniform appearance
    fossils – plants indicate freshwater, shells and corals are marine

SOURCE (or PROVENANCE) an idea of the environment which formed the sediment e.g.
    fine (silt or mud) = still water (lake, deep ocean)
    sand = medium flow (river or ocean current near shore)
    big pebbles = fast mountain stream or storm beach
    big, angular = an agglomerate from landslides or glacier
    limestone = usually a coral reef and may have fossils

Use the textbook or other sources to identify each rock. Write the number of the specimen, the rock's name and then its classification (either clastic or non-clastic) and perhaps two key words for later identification.

EXPERIMENT 6.1 continued

DATA and OBSERVATIONS:

PART A: Describe the rate of settling and draw a table for each fraction (gravel/sand/mud) showing rate of settling against time. Sketch the final layers in the bottle.

PART B: Table of rock properties for each specimen and selected sketches. Copy and complete:

Specimen Number	Overall Colour	Texture (Grains)			Composition	Structures	source	Name
		Size	Shape	Sorting				

Do a sketch of conglomerate <u>and</u> either shale **or** limestone (show any fossils present).

CONCLUSIONS:

In your conclusion for **Part A**, explain why the sediments have settled like this and comment on the time taken for each. Comment on the factors which would control the speed and order of settling

For **Part B**, list the names of the rocks and write two key words to identify the rock and also give a classification for each as clastic (particles can be seen) or non-clastic (no particles e.g. biological or chemical).

Also comment on any internal structures or features seen in each of the rocks and how they and the rock's composition relate to their environment of formation.

EXPERIMENT 6.2 — One Lesson

MODAL ANALYSIS OF SEDIMENT

AIM: To use the technique of Modal Analysis to deduce the nature of a sediment

MATERIALS: Sediment mixture (silt, fine sand, coarse sand, fine gravel), sieve sets, 500 ml plastic measuring beakers, 100 ml measuring cylinders, hand lenses.

BACKGROUND: The nature of the grains in a soil (or sedimentary rock) can indicate the conditions under which the soil (or rock) was formed.

Column graphs of **size** against **percentage of volume** will suggest the most common (**mode**) of size of grain and hence the mode of formation. **Sorting** is shown by the relative percentages graphed for each size.

Sedimentary rocks must first be gently agitated or have their cements dissolved (in acid) before their grains can be tested.

PROCEDURE:

1. Arrange the sieves in stacks with the smallest size on the bottom and the largest on top;

2. In turn, pour 500 ml of the sediment into the top of the sieve set, cover and then shake the combined set vigorously for a few minutes;

3. Separate the sieve set into its separate sieves and measure the volume of each **fraction** (amount) in each sieve;

4. Convert each fraction value to a **percentage volume** of the original volume (e.g. 500 ml) and graph these against the **sieve sizes** as a column graph e.g.

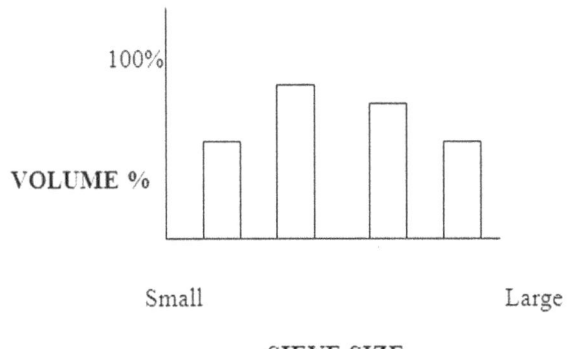

EXPERIMENT 6.2 continued

DATA and OBSERVATIONS:

Draw a full page column graph (as above) which accurately reflects the measurements taken for each sediment sample. Clearly label each with the location of the sand sample.

CONCLUSIONS:

1. Comment on the sorting for each sand sample;

2. Suggest reasons for any differences or similarities between:
    a) beach and dune sands
    b) river gravel and river sand
    c) beach sand and river sand

3. What errors might occur in using volume as a measurement?

4. Can you suggest a more accurate measurement?

5. Comment on the general accuracy and usefulness of this experiment in the study of sedimentary rocks.

Chapter 7: Metamorphic Rocks

EXPERIMENT 7.1   One Lesson

SOME COMMON METAMORPHIC ROCKS

AIM: To examine and describe some common Metamorphic Rocks

MATERIALS: Hand lenses, set of metamorphic rocks (gneiss, hornfels, phyllite, greenstone, slate, marble and schist), and dropper bottles of dilute acid.

PROCEDURE:

1. Carefully examine each specimen with a hand lens.

2. In a table in Results, describe each rock under the following terms:

    colour overall colour

    texture is it foliated or nonfoliated?

    composition if possible describe some minerals seen (revise the practical work on minerals)

    (additional experiment: with care, test the non-foliated rocks with dilute acid. what does this show about the one that reacted?)

    other any other description worth noting.

3. In the results, sketch the gneiss to show its texture.

DATA and OBSERVATIONS:

Draw up a Table of Properties and make sketches of those rocks which show interesting colour or textures e.g.

Rock Number	Overall Colour	Texture	Composition (Minerals)	Other	Name of Rock

CONCLUSIONS:

List the names of each mineral, give two or three key words to help you remember the rock and give the parent rock from which the specimens came by metamorphic action.

CLASSIFICATION OF METAMORPHIC ROCKS

TEXTURE			ROCK NAME	COMPOSITION	PARENT ROCK	METAMORPHIC PROCESS
Foliated	Schistose	Very fine-grained	SLATE	Abundance of dark, flaky and/or prismatic silicae minerals (micas, chlorite, talc, serpentine, hornblende, etc.); quartz	Shale; tuff	INCREASE REGIONAL ↓
Foliated	Schistose	Fine-grained	PHYLLITE		Shale; tuff	
Foliated	Schistose	Medium- to coarse-grained	SCHIST (var. Mica schist, chlorite schist, amphibole schist, etc.)		Shale; intermediate to mafic igneous rocks	
Foliated	Gneissic	Medium- to coarse-grained	GNEISS (var. Garnet gneiss, granite gneis, etc.)	Feldspar abundant; varying amounts of quartz and dark silicate minerals (such as amphiboles, pyroxenes, micas, and garnet)	Felsic to intermediate igneous rocks; arkose; graywacke; mica schist	
Nonfoliated	Granoblastic	Medium- to coarse-grained	METAQUARTZITE	Quartz greatly predominant	Normal and quartzose sandstones	Contact
Nonfoliated	Granoblastic		MARBLE	Calcite and/or dolomite, with or without Ca-Mg silicates	Limestone or dolomite, with or without impurities	Contact
Nonfoliated	Hornfelsic	Fine- to very fine-grained	HORNFELS	Dark silicate minerals predominant	Shale; slate; intermediate to mafic extrusive rocks	Contact
Nonfoliated	Hornfelsic		ANTHRACITE	92 - 98% carbon	Peat, lignite, coal	Contact

ALTERNATIVE EXPERIMENT — One Lesson

(This can be done if the number of specimens or time is limited. It could also be done as an identification practice or a practical examination/assessment.)

SOME COMMON ROCKS

AIM: To examine and describe some common rocks and identify them using a dichotomous key.

MATERIALS: Hand lenses, set of numbered rocks (incl: basalt, sandstone, gabbro, marble, limestone, conglomerate, shale, schist, gneiss and granite), dropper bottles of dilute acid, ruler.

PROCEDURE:

1. Carefully examine each specimen with a hand lens.

2. In a table in Results, describe each rock under the following terms:

    COLOUR overall colour

    TEXTURE any layers, foliation, fossils etc.

    GRAIN SIZE - are there any crystals or sedimentary grains? Estimate sizes if possible using a ruler.

    COMPOSITION if possible describe some minerals seen (revise the practical work on minerals)

    ACID with care, test the non-foliated rocks with dilute acid. What does this show about the chemistry of the one that reacted?)

    OTHER any other observations worth recording.

3. Sketch any specimen showing interesting textures (at least one).

4. Use the key provided on the next page to identify each specimen.

ALTERNATIVE EXPERIMENT continued

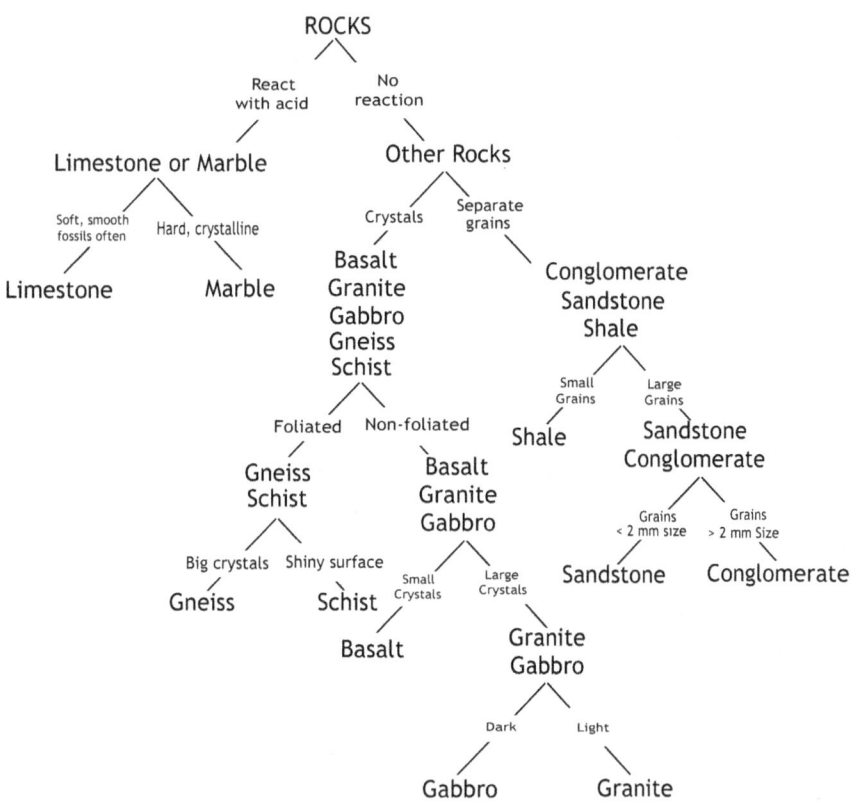

DATA and OBSERVATIONS:

Draw up a table of properties and make sketches of those rocks which show interesting colour or textures e.g.

| Rock Number | Overall Colour | Texture | Composition (Minerals) | Other | Name of Rock |

CONCLUSIONS:

List the names of each mineral, give two or three key words to help you remember the rock and give the parent rock from which the specimens came by metamorphic action.

Chapter 8: Weathering and Erosion

Experiment 8.1 One Lesson

WEATHERING

AIM: To examine and describe causes and features of weathering.

MATERIALS: Limewater, test-tubes & rack, dropper bottles of dilute acid (2M HCl), watch glasses, pieces of limestone, fresh & weathered granite, hand lenses, metal tongs, Bunsen burners and mats

BACKGROUND

Weathering can be physical or chemical. Both cause the breakdown of minerals and rocks; chemical weathering by chemical change and physical weathering by breakdown into smaller pieces.

Limewater (calcium hydroxide solution) is a test solution for carbon dioxide gas. It reacts slowly with CO_2 to form a faint white cloud of calcium carbonate. **CAUTION**: Limewater is also slightly caustic (corrosive to skin).

All **carbonates**, such as calcium carbonate, react with acid to produce CO_2 (and the salt of that acid, e.g. chloride if hydrochloric acid is used)

(in advance: DEMONSTRATION- pour a little limewater into a watch glass and leave .)

PROCEDURE:

PART A: Weathering of Granite (overall)

1. Observe pieces of fresh and weathered granite. Feel both specimen and comment on their overall differences.

2. Carefully examine a <u>fresh</u> piece of granite and identify by sight the main minerals in it (revision from the igneous rocks experiment).

3. Using the hand lens, carefully examine a weathered piece of granite. Try to locate the original minerals (may be changed now) by their relative size, shape and abundance). Describe (colour, hardness, other features) how these minerals have changed.

4. Set out these observations in a Table, sketch and label the <u>weathered</u> granite to a good scale (e.g. half page) and in colour.

PART B: Physical Weathering of Granite

1. WEAR EYE PROTECTION. Carefully heat a small sample of fresh granite by holding it with metal tongs in a blue flame of a Bunsen burner.

EXPERIMENT 8.1 continued

PROCEDURE: continued

2. Quickly drop the hot rock into a glass test-tube of cold water (these two steps may have to be repeated several times). Describe any changes observed.

PART C: Chemical Weathering of Limestone or Marble

1. Place a few pieces of limestone or marble rock into a test tube and CAREFULLY add enough acid to JUST COVER the solid.

2. Test any gas with a drop of limewater hanging on the end of a bacterial loop. What happens? What does this prove?

3. Sketch the main apparatus.

DATA and OBSERVATIONS:

Give descriptions of observations for each part with special attention to the differences between fresh and weathered specimens (e.g. in granite, indicate the minerals and how they have changed). Include WORD EQUATIONS for any chemical changes observed (if possible) and draw neat, well-labelled sketches as appropriate.

CONCLUSIONS:

1. Comment on what was learnt in each of the three sections about weathering.

2. Give word equations for any chemical changes (e.g. minerals in granite, acid on limestone etc.).

3. Relate the artificial, laboratory actions (e.g. using Bunsen burners, adding acid etc.) to the real source of change for the actions observed in this experiment.

4. Discuss any errors which may have prevented the appropriate observations from being made. Suggest any improvements.

EXPERIMENT 8.2 One Lesson

INTRODUCTION TO STREAM TABLE EXPERIMENTS

AIM: To examine how sediment can be eroded, carried and deposited by water.

MATERIALS: Long white trays, small deep trays, retort stands (rod), boss heads and clamps, 500 ml beakers, sand

BACKGROUND:

Sediment is easily eroded by running water and is carried downslope to where it can be deposited. In doing so, the sediment often forms features typical of stream patterns.

PROCEDURE:

1. Set up the following apparatus:

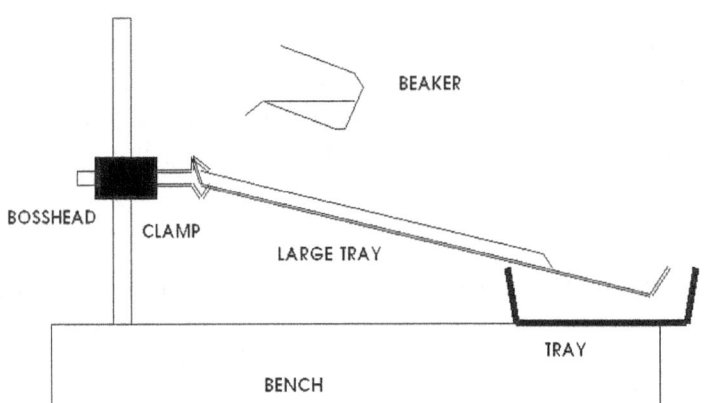

2. Set the tray in an horizontal (level) position (the base of the Bosshead is about 17 cm. above the top of the desk).

3. Add about four beakers full of sand to the tray and moisten until the sand is wet but hard and easy to mould.

4. Smooth the sand out as a flat layer which takes up most of the tray (as shown above).

5. Adjust the Bosshead until the tray is at an angle of about 5 degrees (Bosshead base is about 21 cm from the bench) so that the flat tray slopes and runs into the smaller tray.

6. Calculate the slope of this surface (change in vertical height/change in length).

7. Using the beaker full of water, gently and consistently pour a continual stream of water onto the sand (as shown in the diagram) from about an height of 10 cm.

8. Make general observations as to: shape of the channel; speed of water; and other general effects on the sand.

EXPERIMENT 8.2 continued

PROCEDURE: continued

9. Taking a line of sight (i.e. imagine a straight line from where the water was poured onto the sand and where it flowed out at the end) along the sand surface, measure (in cm.) the maximum width of the valley formed by the stream (several may have to be taken and then an average made).

10. Drain off the water into the collecting tray (or a waste bucket if this is full...**never put sand down the sink**) and reshape the sand as before.

11. Record all general observations about stream activity and see if you can identify any stream features such as meanders, braids and so on (refer to text).

DATA and OBSERVATIONS:

Observe very carefully and record these observations in full sentences in point form. Sketch any shapes or features of interest seen.

CONCLUSIONS:

Write a full conclusion with discussion about the observations made during this experiment.

1. What are the main features of rivers seen in this simulation?

2. What happens to the width of the stream's course as it moves downslope?

3. How is the sediment in the stream carried?

4. Describe the motion of the sediment at the (a)start and (b) end of the river

5. Discuss the use of such models (as in this experiment) in the study of real river environments.

6. Why was the sand moistened before the experiment?

7. What errors could limit the use of such a model? Give some examples from this experiment

8. How such an experiment could be improved.

EXPERIMENT 8.3
Lessons

SOIL TESTING EXPERIMENTS

AIM: To perform some basic experiments on soils.

MATERIALS: Soil samples (local and others if possible), barium sulfate powder, universal indicator in dropper bottles, distilled or de-mineralized water, test tubes (large with stoppers) and test tube racks, clay (whole and powdered), sand, loam, plastic teaspoons, 100 ml measuring cylinders.

BACKGROUND:

The permeability and porosity of soils can be found using experiments as previously described. Other important factors of the soil, such as its organic component and its profile can be found through observation. This activity describes the practical considerations of the acidity (pH) of the soil and its clay content. A good soil should not have too much inert sand (low nutrition and high permeability of water) nor should it have too much clay (low permeability of water). There should also be the right amount of acidity and good topsoil with a high organic component for nutrition.

PROCEDURE:

PART A: Measuring the Acidity (pH) of the soil

1. Place about 1 cm. of soil into a clean test tube.

2. Add about the same amount of barium sulfate powder to flocculate (clump together) any clay particles.

3. Add 10 ml of distilled water.

4. Add several drops of universal indicator and, placing a thumb over the top of the test tube, shake the mixture.

5. Place the test tube in the rack and allow the solids to settle for a few minutes.

6. Compare the colour of the water above the solids to the colour chart for pH.

7. Repeat for several soils if available.

PART B: Clay and Soil - the Farmers' Technique

Having too much clay clogs up the soil and makes water difficult to penetrate; too little and the soil becomes too loose. A quick test for the texture of the soil by farmers on the land is to:

1. Take a small piece of clay onto the palm of the hand and wet it slightly (a farmer might use spittle but here tap water is preferred).

2. Roll it into a small ball.

EXPERIMENT 8.3 continued

PROCEDURE: continued

3. Use the fingers to knead it out into a long ribbon (some farmers roll it out into a long cylinder with the palm of the other hand).

    (Pure clay should give a long, unbroken ribbon or cylinder)

4. Now repeat this with a sample of soil, noting the maximum length obtained until the ribbon or cylinder just starts to break.

    If your ribbon measures less than 2.5 cm long before breaking, you have **loam** or **silt**.
    If your ribbon measures 2.5 to 5.0 cm long before breaking, you have **clay loam**.

PART C: The Effect of Lime on Clay Soils

1. Place about a teaspoon of powdered clay into each of two large test-tubes.

2. Fill both test-tubes with water to about halfway from the top. Stopper and then shake each to disperse the clay.

3. Add limewater (calcium hydroxide solution) to one of the test-tubes, stopper and shake it. Watch what happens carefully.

4. Also stopper and shake the other test-tube and then place both test-tubes into a rack and allow to stand.

PART D: The Components of the Soil

1. Place about 200 ml of soil into a 100 ml measuring cylinder.

2. Fill it up to the 100ml mark, cover the top tightly with the hand and then shake and invert the cylinder.

3. Stand the cylinder upright and allow the mixture to settle. This may take some time (perhaps overnight).

4. When the components have settled (and the water above is clear), measure each component (sand, silt, clay, organics as dark material) as a percentage of the total height of the solid mass in the bottom of the cylinder e.g. if the dark top layer of the organic component is about 2 ml on the scale, and the total mass takes up 200 ml, then the percentage is 2/200 x 100 % = 1.0%.

5. List the components as a table.

6. Repeat with other soils if available.

EXPERIMENT 8.3 continued

DATA and OBSERVATIONS:

PART A: pH Test

1. What is the pH of the soil?

2. What does this means in terms of acidity?
(Answer for each soil tested – if several soils are used give the data in the form of a table).

PART B: Clay and Soil - the Farmers' Technique

1. Describe the nature of the soil sample used.

2. Is this a good test to use? Why?

PART C: Effects of Lime on the Soil

1. What happens when the lime is added to the clay mixture?

2. What effect (if any) will this have on the dry clay soil after treatment?

3. Calcium hydroxide is alkaline. What will this do to the pH of the clay soil? Would this be harmful to plants grown in this soil (in general plants need a pH of about 5.5 to 7.0)?

RESEARCH: (Optional) How can the pH of a very alkaline soil be reduced?

PART D: Soil Components

1. What are the relative percentages of each soil component?

2. How could this soil sample be classified? (see text book)

CONCLUSIONS:

1. Write a full conclusion with discussion about the observations made during this experiment.
2. What are the main factors (a) shown in this experiment and (b) from additional **research** which are important in understanding about the usefulness of soils?

3. What is pH?

4. What is the original source of (a) clay and (b) the organic matter found in some soils?

5. Are there any improvements or additions which should be made in this experiment?

RESEARCH: (Optional)

1. What is the soil type in your local area? Does it need additional improvement to grow crops?

Chapter 9: Environments of Weathering and Erosion

EXPERIMENT 9.1   One Lesson

KARST SIMULATION

<u>AIM</u>: To show a simple model of how limestone caves form below ground.

<u>MATERIALS</u>:   Sugar cubes, sand, clear plastic cups, modelling clay, toothpicks and beakers

<u>BACKGROUND:</u>

Limestone caves systems are formed when water containing dissolved carbon dioxide gas (forming a very weak carbonic acid) percolated through topsoil and into the many cracks in the limestone below. These cracks form when the rock is being compressed and consists of horizontal bedding planes and vertical or oblique fissures. Water seeps down these cracks and the weak acid dissolves out the calcium carbonate, the main constituent of limestone. This takes considerable time so another "rock" (sugar cubes) is used instead of the limestone and water (which dissolves the sugar) is used instead of carbonic acid.

<u>PROCEDURE:</u>

1) Add a small amount of sand (say about 2 cm) to a small transparent plastic cup. This sand represents the rock underlying the "limestone".

2) Stack the sugar cubes so that they are staggered and also hard up against one side of the cup (this will be the viewing portal).

3) Fill more sand around the stacked cubes to represent surrounding sediments.

4) Place a thin layer (about 2-3 mm thick) of modelling clay (plasticine) over the top of the cup.

5) Perforate it with the toothpick to give many holes to simulate the porous topsoil.

6) Using a beaker, slowly trickle water over the top of the model and carefully observe (it may be a fast result). Water can be added until it reaches halfway up the bottom layer of cubes.

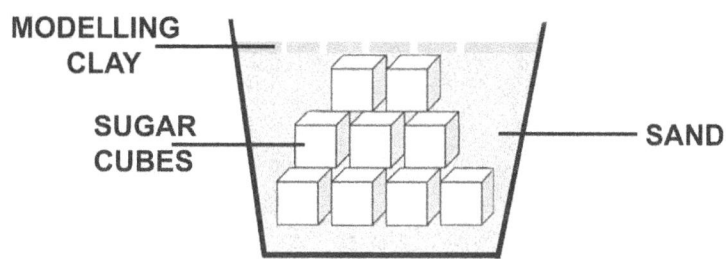

EXPERIMENT 9.1 continued

DATA and OBSERVATIONS:

Observe very carefully and record these observations. Sketch the model after the simulation has been run.

CONCLUSIONS:
Write a full conclusion with discussion about the observations made during this experiment.
1) Where does most of the reaction take place during the initial stages?

2) Is this consistent with what occurs in nature? Why?

3) What happens at the base of the cubes where water has collected (i.e. below the water table of this model)?

4) What errors could limit the use of such a model?

5) How such an experiment could be improved.

RESEARCH: (Optional)
Use the Internet or other resources to find out about limestone caves (a) near your location (b) in other parts where these caves are famous.

1) In some countries such large cave systems are important. Why?

2) What are some of the hazards of living in karst area?

3) What is Spelaeology? Why is it undertaken?

EXPERIMENT 9.2 One Lesson

THE SHAPE OF RIVERS

AIM: To examine the relationship between the slope of the land surface and the shape of a stream.

MATERIALS: Long white trays, small deep trays, retort stands (rod), boss heads and clamps, 500 ml beakers, sand

BACKGROUND:

The shape and pattern of a river system is determined by the gradient of the land surface which also determines the flow rate. The *Sinuosity Index* of a meandering river is the ratio of the actual river length measured along the entire curve of the stream to the down-valley length or straight line between the two points from which the meanders were measured. This sinuosity index has been used to separate single channel rivers into three general classes: straight (SI < 1.05), sinuous (SI 1.05-1.5), and meandering (SI > 1.5).

PROCEDURE:

1. Set up the following apparatus as in Experiment 6.2:

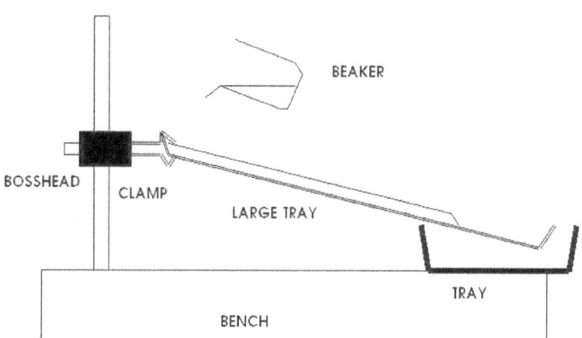

2. Set the large tray to a horizontal (level) position.

3. Add about four beakers full of sand to the tray and moisten until the sand is wet but hard and easy to mould.

4. Smooth the sand out as a flat layer which takes up most of the large tray (as shown above).

5. Adjust the Bosshead so that the tray is at an angle of about 2.5 degrees (Bosshead base is about 21 cm from the bench) so that the flat tray slopes and runs into the smaller tray.

6. Calculate the slope of this surface (change in vertical height/change in length).

7. Using the beaker full of water, gently and consistently pour a continual stream of water onto the sand (as shown in the diagram) from about an height of 10 cm.

8. Make general observations as to: shape of the channel; speed of water; and other general effects on the sand.

EXPERIMENT 9.2 continued

PROCEDURE: continued

9. Measure any meandering sections by measuring the length of the stream (from its centre around its full course) as stream length and also measure the straight line down the stream direction between the two points from which the stream length was measured. This is the course length.

10. Drain off the water into the collecting tray (or a waste bucket if this is full...NEVER PUT SAND DOWN THE SINK) and reshape the sand as before.

11. Repeat Steps 4 to 9 using different angles e.g. 5.0, 7.5, 10 and 12.5 degrees.

12. Record all general observations about stream activity and graph the sinuosity against angle as a line graph (think about graph construction and which axes to use).

DATA and OBSERVATIONS:

Observe very carefully and record these observations. Sketch any shapes or features of interest. Complete a table of angle and Sinuosity Index and then graph this data.

Angle	Stream Length	Course length	Sinuosity Index
2.5^0			
5.0^0			
7.5^0			
10^0			
12.5^0			

CONCLUSIONS:

Write a full conclusion with discussion about the observations made during this experiment.

1. Is there any relationship between the slope of the river bed and its sinuosity? Explain and give reasons.

2. Discuss the use of such models (as in this experiment) in the study of real river environments. Are there other factors which would determine the sinuosity of a river?

3. What errors could limit the use of such a model?

4. How such an experiment could be improved

EXPERIMENT 9.3 — One Lesson

STEREOPAIRS AND LANDFORMS

AIM: To use pairs of stereo photographs (aerial photographs) to interpret features and landforms.

MATERIALS: Pairs of stereo photographs of a variety of landforms and small stereo viewers.

BACKGROUND: Using special stereo cameras (with two lenses at an angle), photographs can be taken from an altitude which will then give a raised 3D image when placed in pairs and view with Stereoscope viewers. The eye sees an impression of raised and sunken features, usually much exaggerated, which often give a better interpretation of a land surface than drawn topographical maps.

PROCEDURE:

1) Use this book or stereo-pairs provided.

2) Open out the legs of the stereo viewer and position it on the page or pairs so that the centre of the viewer is approximately over the join of the paired photograph.

3) Whilst looking down through the viewer and moving it back and forth across the joining line (of the two photos) until the brain suddenly interprets a raised 3D view (sometimes it helps to open and close alternate eyes to get the view with each eye and then move the viewer until both images overlap).

4) Closely examine the 3D features seen and recognise some of the details of each landform.

PHOTO 1:

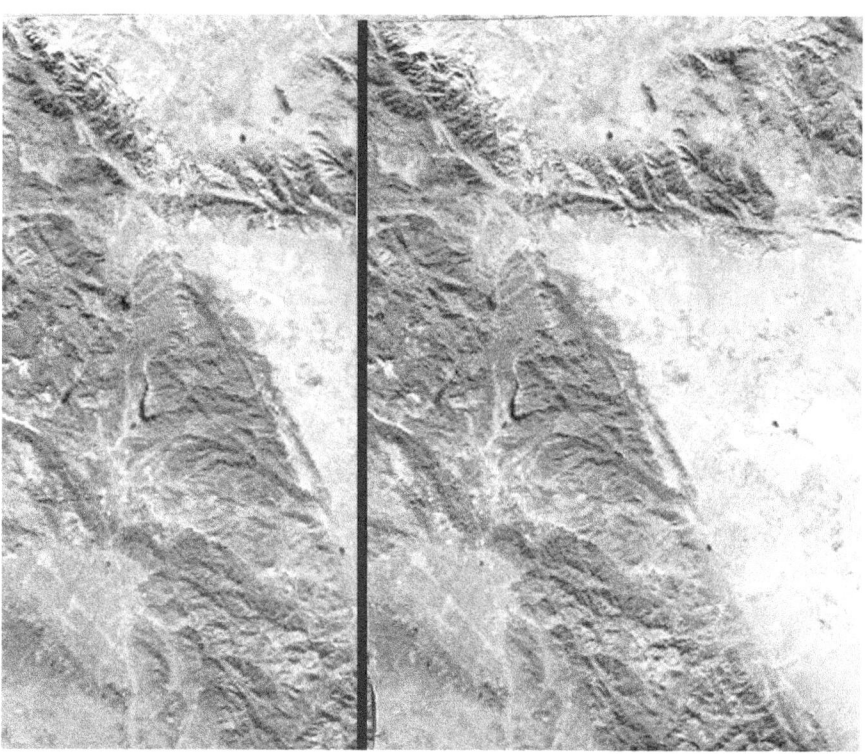

PHOTO 2:

(Photo credit: USGS)

DATA and OBSERVATIONS:

1) What is the main feature or shape seen in each photo?

2) Describe the general land surface of each photo.

3) Suggest a way that these features may have been caused.

CONCLUSIONS:

These photos are low resolution but they can be useful.

1) Comment of the use of satellite and lower altitude aerial photos to locate and describe landforms

2) What errors would limit the use of stereo pairs?

3) How such a technique be improved?

RESEARCH (Optional): Use the Internet to find out how landforms are studied using more modern techniques such as anaglyphs and 3D modelling.

Chapter 10: The Hydrosphere – Waters of the Earth

EXPERIMENT 10.1   **One lesson**

PROPERTIES of WATER

AIM: To explore some of the properties of water.

MATERIALS: plastic rods or rulers, wool cloth, flat dish, paper clips, detergent, droppers or pipettes or drinking straws, waxed paper, glass slides, small beaker, filter paper, alcohol, test tubes, cooking oil, kerosene (paraffin), coloured water,

BACKGROUND: Water covers over 71% of the Earth's surface and is found in all of the spheres of the planet. It is a polar molecule of two hydrogen atoms covalently bonded to an oxygen atom as H_2O. It is polar, that is there is an imbalance of electrical charge at its end with the small hydrogen atoms having their positive nuclei close to the outside of their end of the molecule, and the larger oxygen atom having many negative electrons near the surface of its end of the molecule.

PROCEDURES and OBSERVATIONS:

A. POLARITY of WATER

1. Turn on the tap (faucet) so that there is only a thin stream of water flowing

2. Charge a plastic rod or ruler with a static electrical charge by rubbing it with a piece of woollen cloth or even on the hair. Hair will not work of there is too much hair conditioner or product.

3. Place the charged rod near the water stream.

QUESTIONS:

1. What happens? Why?

2. This experiment will not work well on a humid or rainy day. Why not?

B. SURFACE TENSION

1. Fill a flat dish with water up to its brim.

2. Take a paper clip and carefully lay it onto the surface of water. If this is difficult, bend another paper clip with one arm at right angles to the rest to act as a carrier, lay the second on this 'tray' and then lower it onto the water.

3. Keeping the paper clip where it is, and using a dropper or pipette, carefully add more water drop by drop to overfill the container until the water bulges over the sides. Note the shape of the top of the water surface.

4. Now add a small drop of detergent to the water.

EXPERIMENT 10.1 continued

QUESTIONS:

1. What happens to the paper clip when placed onto the top of the water? Why?

2. What is the shape of the water surface when the dish is overfilled? What name is given to this curved surface?

3. What happens to the paper clip when the detergent is added? Why?

C. COHESIAN and ADHESION

1. Place a smooth piece of waxed paper on the desk and flatten it out.

2. Using a dropper or pipette, place one drop of water onto the waxed paper.

3. Slowly tilt the piece of paper until the drop moves across the surface.

4. Repeat the last two steps using a clean, dry glass slide.

5. Place a length of glass tubing (e.g. the end of the pipette) or a transparent drinking straw into a beaker of water and carefully look at the water level in the tube.

QUESTIONS:

1. Comment on the movement of the water drop down the surfaces of the waxed paper and the glass. Why are they different?

2. Where is the top of the water level in the glass tube? Explain.

D. CAPILIARY ACTION - another version of Part C

Take a strip of filter or blotting paper and use a marking or felt pen to make a 1-2 mm spot about 2-3 cm from one end of the strip.

Hold this strip very still so that the end with the spot is in water with the spot about 1 cm above the level of the water. This is a technique similar to Paper Chromatography.

Repeat this activity using different inks or dyes.

QUESTIONS:

1. What happens to the water contacting the strip? Why? (think of Part C)

2. What happens to the ink spot? Explain any observations.

EXPERIMENT 10.1 continued

E. HEAT OF VAPORIZATION

Place a drop of water onto the finger and then gently bow onto it.

Repeat using a small amount of alcohol.

QUESTIONS:

What happens to the temperature of the tip of the finger? Why?

Is there any difference using the alcohol? If so explain?

F. SOLUBILITY and MISCIBILITY (ability to mix)

1. Pour coloured water into a test tube to about one third volume.

2. Add about the same amount of cooking oil.

3. Holding the thumb securely over the top of the test tube, invert it and shake vigorously.

4. Return the test tube to the upright position and observe.

5. Add a small amount of alcohol to this mixture and shake as before.

QUESTIONS:

Describe what happens in the first part of this activity. Is water miscible or immiscible with the cooking oil? Explain.

Comment on what happens when the alcohol is added. What does this say about the alcohol molecule?

CONCLUSIONS:

Summarise these properties of water in a table and give a reason for each of these properties in terms of the water molecule e.g.

PROPERTY	REASON

RESEARCH (Optional)

1. How is capillary action used in living things?

2. What is Paper Chromatography? How is it used as a scientific tool?

EXPERIMENT 10.2
MAPPING THE DEPTHS
One lesson

AIM: To use depth recording data to construct a typical profile across an ocean.

MATERIALS: Data provided, graph paper or plain paper ruled, pencils

BACKGROUND:

The use of a technique called sonar (SOund Navigation And Ranging) greatly improved the ability of ships to measure the depths of the ocean (Bathymetry). In SONAR a ship sends out a pulse of sound (a ping), which is reflected by the sea floor and detected with the ship's instruments. If the time it takes for a reflected ping to be heard is measured and the speed of sound in water is known, then the water depth can be derived. Using sonar, ships could record continuous depth measurements without stopping.

In this exercise, an oceanographic ship sails eastward from Puerto Rodrigo to Orupenda Bay, taking SONAR readings underway at various intervals at stations marked on the map. North is at the top of the page.

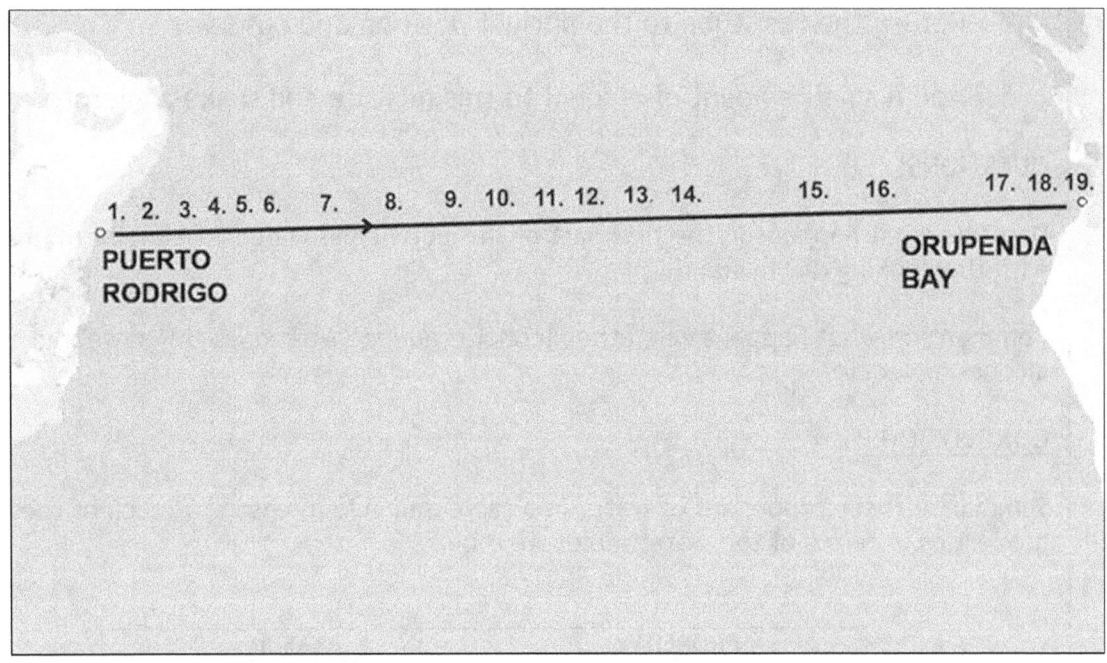

The total time for a reflection to return to the ship was recorded at each station and is given in a table on the next page:

EXPERIMENT 10.2 continued

STATION	DISTANCE TRAVELLED (km)	REFLECTION TIME (seconds)
1. Puerto R.	0	0
2.	100	0.3
3.	200	4.0
4.	280	5.3
5,	400	6.4
6.	600	6.9
7.	1000	6.6
8.	1400	6.3
9.	1800	5.8
10.	2000	5.1
11.	2100	3.5
12.	2200	3.3
13.	2400	5.8
14.	2600	6.6
15.	3200	6.9
16.	3600	5.9
17.	3800	4.2
18.	3900	0.2
19. Orupendo	4000	0

The depth of the ocean at any station can be found by the formula:

$$D = V \times T/2$$

Where D = depth in metres,
T = the total time of reflected wave and
V is the average velocity of sound in water which is 1507 metres/second.

PROCEDURE:

1. Redraw this table adding a fourth column for depth.

2. Use the formula above to calculate the depth (in metres) of the ocean at each station.

3. On a sheet of A4 graph paper or suitable plain paper ruled up, draw a large graph with DEPTH in metres along the vertical axis and the distance travelled in kilometres along the horizontal axis.

4. Mark in each station and join the plotted points as a curve of best fit and show sea level along the top of the graph.

EXPERIMENT 10.2 continued

DATA and OBSERVATIONS:

Redraw the table adding the fourth column for depth.
Draw the graph using an appropriate (separate) scale for depth and distance travelled. This graph should be drawn with the A4 paper turned on its side and will give a shape which will have a very large vertical exaggeration and thus greatly exaggerated slopes.

CONCLUSIONS:

1. What is the main feature seen in this cross-section of the ocean floor? How was it formed?

2. What other oceanographic methods could be used to confirm the nature of this feature?

3. What name is given for the section of this cross-section:

    a. just offshore east from Puerto Rodrigo and just offshore west of Orupenda Bay?
    b. between Stations 2 to 3?
    c. between Stations 4 to 6?
    d. between Stations 6 to 8 and 14 to 15?

4. What is the slope of the section between:

    a. Stations 2 to 3? (Hint: Slope = Rise/Run and use trigonometry)? And
    b. Stations 4 to 6?

5. What is the general depth of this section in 2 (c) above?

EXPERIMENT 10.3   Two or three lessons

SOME WATER SAMPLING TECHNIQUES

AIM: To test some of the properties of water using some simple environmental water testing techniques.

MATERIALS: Pond or river water samples collected by students or teacher at local places and professional water sampling kit (preferred). Alternatively, simple laboratory equipment may be used to show the basics of such kits using prepared water samples and laboratory equipment including: buckets with turbid water, class-made Secchi Disks (white cardboard, permanent pens, drawing compass, string, mass carriers from Physics), universal indicator, test-tubes, pH colour charts, indigo carmine dye, potassium hydroxide, dextrose, ammonium molybdate solution, phosphated water, conc. nitric acid, Bunsen burner and mat, agar plates, copper wire and loop, incubator.

BACKGROUND:

Environmental Earth Scientists often spend considerable time in the field sampling aspects of the environment for mineral wealth or to estimate the health of the environment. Usually they will walk upstream taking samples at specific points, especially where one stream joins another usually with several streams being sampled along the stream traverse. In other research activities, a specific waterway such as a pond or lake or even sea may be sampled to determine the health of the environment, especially if some pollution to the waterway is suspected. These water samples may be tested simply in the field as a convenience or they may be taken back to the testing laboratory if the research is long and extensive.

PROCEDURE:

Work through the procedures of the water sampling kit and test the water sample(s) provided.

PART A: Turbidity

This is a measure of the cloudiness of the water. It is usually measured in the field, but with the appropriate sample, it can be done as an example in the laboratory. A visual reference pattern called a Secchi Dish is used.

1. Make a Secchi Disk pattern by drawing a 20 cm diameter circle onto stiff, thick cardboard, quartering it and blacking in the quadrants with a permanent pen. Push a hole in its centre and insert this over a mass carrier (from the Physics lab) or attach a large washer to underneath the base. Attach string to the upper end of the mass carrier (or through the hole and washer at base).

2. Lower this into a bucket of turbid water and hold the string at water level when the disk just disappears from view.

3. Holding the string at the water level mark, pull out the disk and measure the length in centimetres from the base of the disk to the point on the string representing the depth at which the disk just disappeared. This is a measure of the clarity of the water.

Retain these disks for later field testing of local water.

EXPERIMENT 10.3 continued

PROCEDURE: continued

PART B: pH

This is a measure of the acidity or alkalinity of the water. High values or acidity or alkalinity could indicate the presence of pollution or an imbalance of the ecology. Organisms have different pH tolerances but most seem to be around that of neutral conditions at pH = 7 (+/- 1.0). Some extremophiles can live in conditions well outside of normal values.

1. A small sample of the water to be tested is placed into a test tube and a few drops of Universal Indicator are added.

2. Shake the test tube and compare the pH to a standard pH colour chart.
(or see https://www.rapidonline.com/edu-resources/docs/ph-scale.pdf)

3. Record this value in a table for the whole water test under pH.

PART C: Temperature

This is a very simple and obvious test as some waters become polluted with heat from industrial plants and the like or undergo extreme climatic changes.

1. Leave a bucket of water outside in direct sunlight to simulate a small pond during an extremely hot summer. Retrieve the buckets and carefully hold the thermometer about 1 cm below the surface and measure the temperatures.

2. Repeat this at different depths such as 5 cm, 10, cm and 20 cm.

3. Draw up a table showing these depths and temperatures.

PART D: Oxygen Content

The teacher should make up an Indigo Carmine solution just prior to the experiment. This consists of 0.25 g of indigo carmine dye added to 2.5 g of dextrose (or glucose) in 220 ml of distilled water. To this is added 280 ml of glycerol (glycerine) and the mixture (solution A) stored in the refrigerator (it will last for a week). Another solution (solution B) is made by dissolving 100 g of potassium hydroxide in about 250 ml of water. Just before the experiment, 20 ml of solution A is mixed with 5 ml of solution B – this will settle down from a red colour to the standard yellow.

Add 4-5 ml of the test mixed solution to about 300 ml of water to be tested.

Compare this to the colours below to find the oxygen level in parts per million (ppm):

Yellow	0.000
Orange	0.005
Orange-Pink	0.010
Pink	0.015
Pink-Red	0.025
Red-Purple	0.050
Purple	0.100

EXPERIMENT 10.3 continued

PROCEDURE: continued

As a test of calibration, test a sample of water which has been boiled strongly (which removes all of the dissolved oxygen) and then been covered and allowed to cool. It should give a zero reading.

(also see: https://www.youtube.com/watch?v=_e8ENtdBmlc)

PART E: Phosphorus as Orthophosphate ions (Qualitative Method only)

As a bad sample of water contaminated with agricultural phosphate, the teacher may add some soluble phosphate salt (e.g. potassium phosphate) or a little phosphoric acid (often found in metal cleaners) or even a little amount of phosphate fertilizer.

1. Add a small amount of the sample to a test tube (about one third full) and acidified it with a few drops of conc. nitric acid CARE!).
2. Add a few drops of ammonium molybdate solution (saturated) and the test tube gently heated.

Phosphate ions are indicated by the bright yellow precipitate of ammonium phosphomolybdate.

PART F: Nitrates - the Brown Ring Test

Nitrates are all very soluble and get into natural waters by a number of methods, often as a result of agricultural leaching. The teacher can make up a test solution using any of the common nitrates such as sodium or ammonium nitrate.

1. Add a small amount of test solution to a test tube and add some iron III sulfate solution (saturated) so that the test tube is only half filled.
2. Slowly add drops of conc. sulfuric acid (CARE) to the test tube which is slightly tilted so that the denser acid will slide down the inside and go below the solution of the nitrate.
3. A brown ring will form at the junction of the two layers, indicating the presence of the nitrate ion. This test is sensitive up to a concentration of 1 in 25,000 parts.

PART G: Testing for Bacteria

At least two days before the experiment, the teacher should make up a sample of bacterial water by placing some plant leaves with roots and soil into a 500 ml beaker of warm water and set it aside in a warm, dark place.

1. Take a previously made agar plate consisting of agar gel in a petrie dish and place it on the desk near a sample of the bacterial water in a small beaker.

2. Obtain a length of copper wire which has had a small loop made in one end.

EXPERIMENT 10.3 continued

PROCEDURE: continued

3. Sterilize the loop in the blue flame of a Bunsen burner.

4. Cool it and dip it into the water specimen near any decayed vegetation or soil.

5. Carefully open one end of the Petrie Dish to a narrow slit and quickly make several passes across the agar with the contaminated loop. Make a crisscross pattern with these passes.

6. Quickly close the petrie dish, seal with transparent tape.

7. Turn it upside down to make an air-tight seal and place into a warm incubator for several days (say 4). Observe and record the plates each day.

DATA and OBSERVATIONS:

For each of Parts A to G, briefly describe the test and the result. If needed draw a sketch of the finished result and the apparatus e.g. the circular petrie dish and the pattern of bacteria on it.

CONCLUSIONS:

1. List the type of tests and the results of each as a summary of the experiment.
2. For each part, describe any error(s) which could limit the efficiency of the test.
3. Comment on the use of specially-prepared field kit over the laboratory methods indicated above.
4. Explain why environmental Earth Scientists would be interested in such tests.

Research (optional):

1. Geological field work also involves stream traverses and water sampling looking for sources of valuable minerals such as lead, zinc and other metals. What chemical tests would show the presence of these metals?
2. Sulfate and chloride ions are often present in water. What chemical tests can be used to detect these ions? They are relatively simple and can be done in the laboratory. Try testing tap water for these two ions.

Chapter 11: The Atmosphere – The Air Above

> **EXPERIMENT 11.1** One lesson
>
> **RELATIVE HUMIDITY**
>
> <u>AIM:</u>   To construct a simple wet/dry bulb hygrometer and then measure the relative humidity of the air.
>
> <u>MATERIALS:</u>   Thermometers (two/group), cotton gauze (or open weave cloth), small beaker or plastic container,   stand with support, clock/stopwatch
>
> <u>BACKGROUND:</u>
>
> Relative humidity is the amount of water that the air can contain at that given temperature and is measured as a percentage. Evaporation rate depends on this amount with higher evaporation occurring at lower relative humidity. As water evaporates from material, it takes some of the heat from the material and drops its temperature. The amount that the temperature drops can be used to determine the relative humidity of the surrounding air.
>
> <u>PROCEDURE:</u>
>
> 1. Take two thermometers and wrap a small amount of material gauze around the bulb of one of them.
>
> 2. Hang both thermometers from the support so that there is some separation between the two.
>
> 3. Wet the bulb with the gauze and hang the end of the gauze into the container now filled with water.
>
>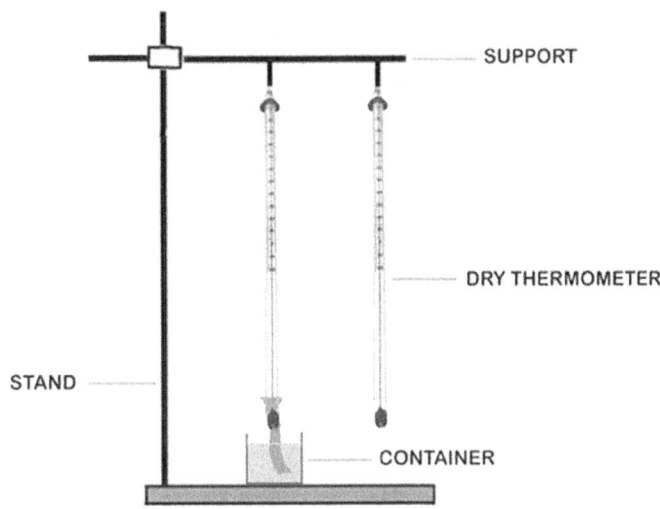

EXPERIMENT 11.1 continued

PROCEDURE: continued.

4. Stand the apparatus in a place with a free air circulation. If the classroom is air conditioned, continue with the experiment but then repeat it outdoors in the corridor or a shaded area.

5. Measure the temperatures of both thermometers every minute until there is no change in the thermometer with the wet bulb.

6. Taking the lowest temperature of the wet bulb and the difference between this and the temperature of the dry bulb, consult the following table to determine the relative humidity.

7. Leave the apparatus until near the end of the lesson (in the classroom) and re-calculate the relative humidity using the latest values shown on the thermometers. Is there a change?

DATA and OBSERVATIONS:

Set out a table showing Time, Dry Bulb Temperature and Wet Bulb Temperature.

Also show the temperatures and the relative humidity inside the classroom at the start of the experiment and near the end of the lesson. If appropriate show the temperatures and relative humidity outside and in the air conditioned classroom.

Wet-and-Dry Bulb Thermometer Relative Humidity

Dry Bulb Temp.	Dry Bulb Temperature minus Wet Bulb Temperature (zero difference =100% relative humidity)													
	1°C	2°C	3°C	4°C	5°C	6°C	7°C	8°C	9°C	10°C	12°C	14°C	16°C	18°C
10°C	88%	77%	66%	55%	44%	34%	24%	15%	6%					
11°C	89%	78%	67%	56%	46%	36%	27%	18%	9%					
12°C	89%	78%	68%	58%	48%	39%	29%	21%	12%					
13°C	89%	79%	69%	59%	50%	41%	32%	22%	15%	7%				
14°C	90%	79%	70%	60%	51%	42%	34%	26%	18%	10%				
15°C	90%	80%	71%	61%	53%	44%	36%	27%	20%	13%				
16°C	90%	81%	71%	63%	54%	46%	38%	30%	23%	15%				
17°C	90%	81%	72%	64%	55%	47%	40%	32%	25%	18%				
18°C	91%	82%	73%	65%	57%	49%	41%	34%	27%	20%	6%			
19°C	91%	82%	74%	65%	58%	50%	43%	36%	29%	22%	10%			
20°C	91%	83%	74%	66%	59%	51%	44%	37%	31%	24%	11%			
21°C	91%	83%	75%	67%	60%	53%	46%	39%	32%	26%	15%			
22°C	92%	83%	76%	68%	61%	54%	47%	40%	34%	28%	16%	5%		
23°C	92%	84%	76%	69%	62%	55%	48%	42%	36%	30%	19%	7%		
24°C	92%	84%	77%	69%	62%	56%	49%	43%	37%	31%	20%	9%		
25°C	92%	84%	77%	70%	63%	57%	50%	44%	39%	33%	22%	13%		
26°C	92%	85%	78%	71%	64%	58%	51%	46%	40%	34%	23%	14%	4%	
27°C	92%	85%	78%	71%	65%	58%	52%	47%	41%	36%	26%	16%	7%	
28°C	93%	85%	78%	72%	65%	59%	53%	48%	42%	37%	27%	17%	8%	
29°C	93%	86%	79%	72%	66%	60%	54%	49%	43%	38%	29%	20%	11%	
30°C	93%	86%	79%	73%	67%	61%	55%	50%	44%	39%	30%	20%	12%	4%
32°C	93%	86%	80%	74%	68%	62%	56%	51%	46%	41%	32%	23%	15%	8%
34°C	93%	87%	81%	75%	69%	63%	58%	53%	48%	43%	34%	26%	18%	11%
36°C	93%	87%	81%	75%	70%	64%	59%	54%	50%	45%	26%	28%	21%	14%
38°C	94%	88%	82%	76%	71%	65%	60%	56%	51%	47%	38%	31%	23%	17%

For example: The relative humidity with dry = 24°C and wet = 20°C is 69% at 24°C

EXPERIMENT 11.1 continued

CONCLUSIONS:

1) What is the relative humidity:

   (a) at the start of the lesson?
   (b) near the end of the lesson?

2) If there was any change, explain why this may have occurred?

3) If appropriate, was there any difference in humidity between an air conditioned room and outside? If so, why?

4) Explain why it is better to do the washing on a day with low relative humidity?

Research (Optional):

Use the Internet and class discussion to consider what household or industrial processes which require low relative humidity.

EXPERIMENT 11.2 — THE ANEROID BAROMETER

One lesson

AIM: To construct a simple Aneroid Barometer and note any changes in air pressure.

MATERIALS: Large, open-mouth jars, rubber balloons or thin sheet rubber, rubber bands, drinking straws, glue, stands, ruler

BACKGROUND:

Aneroid (without air) barometers work because the air outside changes whereas the small amount of air inside the box of the barometer does not. As the air changes outside, the relative air pressure between that inside the box of the barometer and that outside will cause a movement of the pointer of the barometer (see text).

PROCEDURE:

1) Obtain a large, open-mouthed jar which has a uniform rim.

2) Stretch some sheet rubber over the top (or use a rubber balloon) and secure this strongly with a rubber band and/or tape.

3) Glue the end of a drinking straw to the centre of the stretched skin and stand the apparatus so that the other end of the drinking straw is against a supported scale (e.g. a ruler)

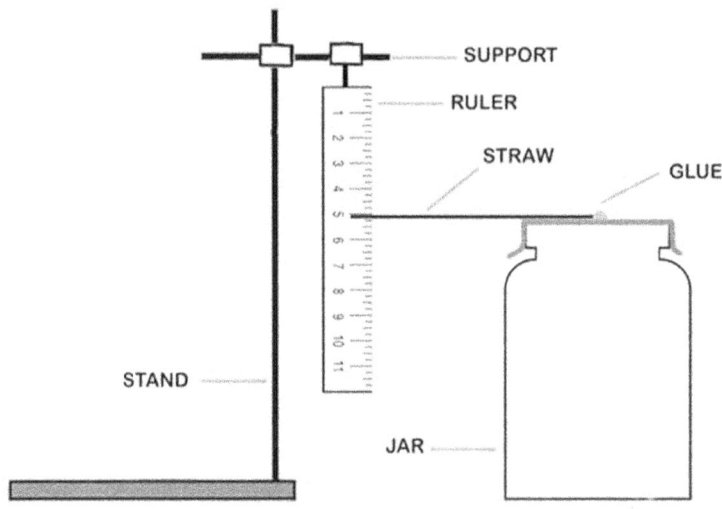

4) Measure any change in the position of the end of the drinking straw (and hence the relative air pressures between the air inside the jar and that outside) over several days.

EXPERIMENT 11.2 continued

DATA and OBSERVATIONS:

Describe what has happened over several days to the end position of the drinking straw.

CONCLUSIONS:

1. Did the relative air pressure change over several days? How?

2. What would be the errors in this experiment or the disadvantages of such a device for measuring air pressure?

3. How could this crude instrument be calibrated?

4. What would be the advantage of having a jar with less air pressure inside (as in the box of a real Aneroid Barometer)?

Research (Optional): Use the Internet to find out more about barometers, comparing the usefulness of Aneroid Barometers against Fortin Barometers.

EXPERIMENT 11.3

One Week

WEATHER MONITORING

AIM: To monitor the local weather for one week (or longer), noting any changes in temperature, relative humidity, air pressure, wind speed and direction and cloud cover and type.

MATERIALS: Thermometers, hygrometers, barometers (commercial brands or as made)

BACKGROUND:

Weather is a study of atmospheric changes using various instruments, computer simulations and observation to make predictions about future conditions for a range of commercial, industrial and personal reasons. Most scientific institutions, ships at sea and farmers keep detailed observations and records about the weather to make predictions about the weather.

PROCEDURE:

1. Set up a 'weather station' of the main instruments list above, in a secure place under cover and away from direct sunlight. Some places have a proper Stevenson's Screen or electronic weather station equipped for regular weather measurement.

2. Cloud cover can be estimated by halving sections of the sky several times to get an eighth part to get an idea of the clouds of the cloud cover in fractions of eights e.g. 8/8 is completely covered, 4/8 is only half covered. Also note the type of cloud using the diagrams in the textbook.

3. Wind direction can be found by finding geographical north (say with a compass or the Sun's position) and then looking at the direction that a flag or trees are moving. A small handkerchief held in the hand would also be satisfactory. Give the direction that the wind is coming from as the wind direction. Wind speed can be subjective e.g. no wind. Light wind. Gentle breeze, strong wind etc.

4. Over the week, measure the parameters of the weather either in class time, or at a set period of day. This may be done on an individual or class group roster. Plot these parameters onto a table/chart.

5. After the week has ended, draw graphs of temperature, relative humidity and pressure then extrapolate (i.e. extend the end of the graph plot) these graphs to predict what the weather will be like in the next few days.

DATA and OBSERVATIONS:

Use a table to record each parameter daily.

If measured in proper units using scientific versions of instruments, complete graphs for temperature, relative humidity and air pressure.

EXPERIMENT 11.3 continued

CONCLUSIONS:

1. Has the weather changed over the week? How (give details)

2. Was the weather in the next week as predicted? Comment.

3. List all of the problems or errors which could occur with such short-term weather measurement.

RESEARCH (Optional):

1. Where is the place where the weather is measured and recorded locally?

2. How is weather data gathered on a wider scale?

3. How are weather predictions made?

Chapter 12: The Biosphere – Life on Earth

EXPERIMENT 12.1   One Lesson

INTRODUCTION to the BIOLOGICAL MICROSCOPE

AIM: To become familiar with the compound biological microscope.

MATERIALS: Biological microscope, ruler, scissors

BACKGROUND:

Modern biological microscopes are precision scientific instruments and as such must be handled with care. They differ from petrological microscopes in that they are usually of a higher power, have a fixed not rotating stage and usually do not have polarizing filters, although there are special versions which do. There are several important procedures for using these microscopes:

1. Do not touch the faces of the lenses or the mirror. Use special lens cleaning cloth to clean them.

2. Always focus upwards to prevent grinding the objective lenses into the specimen. Whilst looking at the whole microscope (i.e. the eye is not looking through the eyepiece) use the coarse focusing knob to move the objective down to just above the specimen, see which way the focusing knob moves the barrel and then put the eye to the eyepiece and FOCUS UPWARD.

3. Always raise the body tube when changing the objective lenses.

4. Take care when carrying the microscope – one hand holding the body and the other under the base.

5. Adjust the angle of the mirror and the diaphragm under the stage to give the best light. It is best to keep the microscope upright even though some can be angled, this way wet specimens do not slide off the stage.

6. Use a cover slip over the specimen on the glass slide if possible.

PROCEDURE:

1. Set up the microscope in a stable position to give good light to the mirror.

2. Carefully examine the parts of the microscope from the diagram given on the next page.

3. Move the coarse focusing knob to see how the objectives move up and down.

4. Move the eyepiece body around to get the lowest magnification eyepiece – this is found by looking at the numbers on the barrels of the eyepieces e.g. x 10, x 20, x 40 etc. The lowest magnification eyepiece is also the shortest in length.

EXPERIMENT 12.1 continued

PROCEDURE: continued

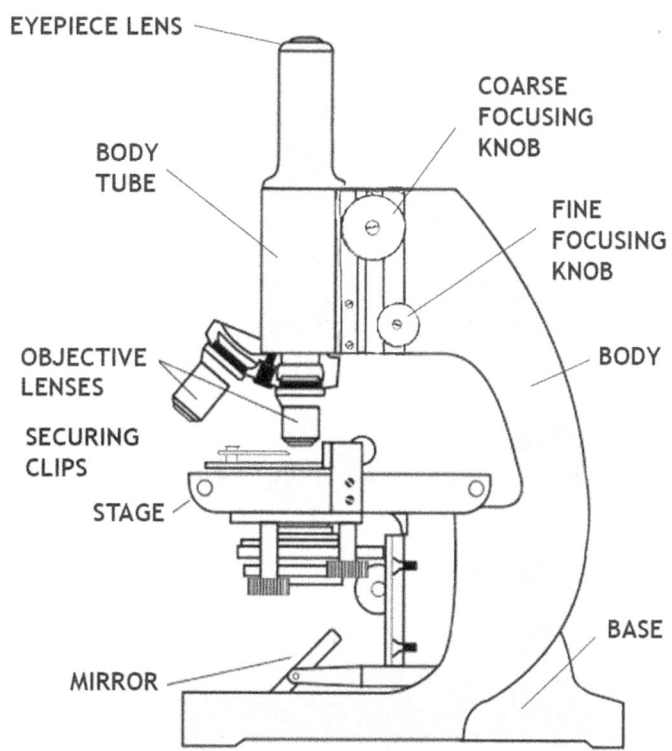

5. Place a transparent ruler onto the stage and move the eyepieces down to it and focus upwards until a clear image is found (students wearing glasses may remove them for comfort).

6. Use the scale on the ruler to estimate the diameter of the field of view (the circle containing the image).

7. Repeat the last two steps to estimate the diameter of the field of view for all objective lenses. Compare the amount of light and the ability to get a clear focus for each view of the ruler using different objective lenses.

8. Draw up a table showing magnification and its field of view diameter. The total magnification is found by multiplying the magnification shown on the eyepiece with that of the objective lens e.g. with an eyepiece at x 10 and an objective at x 20, the total magnification is x 200.

9. Take a human hair and carefully place it onto a glass slide (perhaps stuck with a little water) and observe its features using the low power objective. Estimate the width of the hair. Repeat using the next higher power objective.

EXPERIMENT 12.1 continued

PROCEDURE: continued

10. Move the glass slide to one side and see how its image moves in the field of view.

11. Draw a circle on the paper of your report and sketch the hair in the field of view. Microscope sketches are best down from memory, NOT with one eye looking into the eyepiece whilst drawing. Carefully examine the specimen, then turn away and draw the sketch. Repeat as necessary.

12. Give the scale of the total magnification (e.g. x 100) and label the sketch.

13. The last three steps may be repeated using different coloured hair to see if there is any difference between colour and width of hair.

DATA and OBSERVATIONS:

Draw up a table showing the field of view diameter in millimetres and its corresponding total magnification (i.e. objective x eyepiece values)

Draw a circle about 8-10 cm diameter and sketch in it the view of the hair(s) seen under low power. Give the magnification and any appropriate label.

CONCLUSIONS:

1. What happens to the light intensity and clarity of image as the magnification is increased?

2. What is the diameter of the human hair examined?

3. Compare the movement of the glass slide with the observed movement of its image in the field of view.

4. Is there any difference with hair of a different colour?

EXPERIMENT 12.2 One Lesson

EXAMINATION of a PLANT CELL - ONION EPIDERMAL CELLS

AIM: To prepare, mount, examine and draw a typical plant cell taken from the outer tissue (epidermal tissue) of an onion.

MATERIALS: Biological microscope, glass slides, cover slips, Tincture of Iodine with dropper (CARE: this will stain skin and clothing - some people are also highly allergic to iodine. Do not touch it!).

BACKGROUND:

Onion epidermal (outer layer) tissue provides excellent cells to study as the main cell structures (organelles) are easy to see when viewed with the microscope at medium power. Try to find the large circular nucleus in each cell containing the nucleoli. Also locate the well-developed cell wall and the cell membrane just within it. There may also be a few vacuoles.

PROCEDURE:

(Also see: https://www.youtube.com/watch?v=eD1CdfRycqs)

1. Obtain a clean and dry glass slide and cover slip.

2. Remove a single thin layer of epidermal cells from the inner (concave) side of the peeled onions - the thinner the better.

3. Place a small piece of this onto a glass slide ensuring that it does not become folded over. Place a drop of iodine stain onto the onion tissue.

4. Put the cover slip on the stained tissue by placing an edge of the cover slip onto the glass slide near the specimen and allowing it to gently fall on top of it. Gently tap out any air bubbles using a pencil or end of a pen.

5. Set the objective lens to lowest power and watch whilst it is brought down to just above the cover slip.

6. Carefully focus upward until a clear view is seen. Try to identify the main parts of the plant cell.

7. Repeat the last two steps using medium and then high power objectives until the best view of a single cell is found. The slide may have to be moved about also.

8. Sketch a diagram of the plant cell and give the magnification and add appropriate labels.

EXPERIMENT 12.2

DATA and OBSERVATIONS:

Draw a circle about 8-10 cm diameter and sketch in it the view of the onion cell(s) seen under the best power. Give the magnification and any appropriate labels.

CONCLUSIONS:

1. Why was iodine used?

2. What distinguishes the nucleus from any other organelle observed?

3. Comment on the general shape of the cells and how they are packed together. What factors determine the size and shapes of these cells?

EXPERIMENT 12.3
One Lesson

EXAMINATION of an ANIMAL CELL - HUMAN CHEEK EPIDERMAL CELLS

AIM: To prepare, mount, examine and draw a typical animal cell taken from the outer tissue (epidermal tissue) of the inside of the human cheek

MATERIALS: Biological microscope, glass slides, cover slips, methylene blue dye with dropper (CARE: this will stain skin and clothing – Do not touch it!), 1% saline solution (1% salt in water), paper towel.

BACKGROUND:

The best place to safely obtain some typical animal cells is from the inside of the cheek from inside of the mouth. Care must be taken to use fresh cotton buds (toothpick as a poor alternative) to scrape the inside of the check. Do NOT use the buds used by others and dispose of it into an appropriate waste container (provided). The main cell structures (organelles) should be able to be viewed with the microscope at medium power. Try to find the large circular nucleus in each cell containing the nucleoli. Also locate the cell membrane and the cytoplasm.

PROCEDURE:

(See also : https://www.youtube.com/watch?v=C6-Nat8fwZw)

1. Obtain a clean and dry glass slide and cover slip.

2. Use a cotton bud to smear some of the saline solution onto the centre of the glass slide.

3. Using a CLEAN, FRESH cotton bud (or toothpick), lightly scrape the inside of the cheek.

4. Scrape this onto the centre of the glass slide washed with saline.

5. Put the cover slip onto the smear by placing an edge of the cover slip onto the glass slide near the smear and allowing it to gently fall on top of it. Gently tap out any air bubbles using a pencil or end of a pen.

6. Place the end of the dropper of methylene blue gently onto the edge of the cover slip and squeeze the smallest amount of dye out so that it runs under the cover slip and across the smear. Blot any excess with paper towel.

7. Set the objective lens to lowest power and watch whilst it is brought down to just above the cover slip.

EXPERIMENT 12.3

PROCEDURE: continued

8. Carefully focus upward until a clear view is seen. Try to identify the main parts of the cell and how cells are positioned.

9. Repeat the last two steps using medium and then high power objectives until the best view of a single cell is found. The slide may have to be moved about also.

10. Sketch a diagram of the cheek cell and give the magnification and add appropriate labels.

DATA and OBSERVATIONS:

Draw a circle about 8-10 cm diameter and sketch in it the view of the cheek cell(s) seen under the best power. Give the magnification and any appropriate labels.

CONCLUSIONS:

1. Why was 1% saline solution used to cover the glass slide?

2. Why was methylene blue used?

3. What distinguishes the nucleus from any other organelle observed?

4. Comment on the general shape of the cells and how they are positioned in the field of view.

5. Compare and contrast the size, shape and features seen within these cheek cells from those of the plant cells seen in the previous experiment.

EXPERIMENT 12.4   One Lesson

CLASSIFICATION OF ORGANISMS

AIM: To the key charts from the textbook to classify organisms seen as preserved specimen or from photographs.

MATERIALS: Preserved/embedded specimens or photos/slides

BACKGROUND:

In order to systematically study living organisms, biologists have devised method of classification using the binomial system of Linnaeus. These methods usually involve the outside body features and some of the feeding habits of the organism. As time progressed and new research discovered new organisms or new functions of previously-classified organisms, systems of classifications often also change to give the best ways of distinguishing one organism from another.

PROCEDURE:

1. Draw up a chart showing the number and/or name of the specimen studied or view down the left hand column and its Kingdom Division/Phylum and other classification groups along the horizontal axis.

2. Select or view each specimen in turn and decide firstly to which of the main Kingdoms that organism belongs using the key chart of the textbook.

3. Next, continue using the key chart for that particular kingdom (usually Plant or Animal) to continue classifying that specific organism.

4. Discuss in the group how each specimen was classified.

DATA and OBSERVATIONS:

Chart of classification and some sample sketches (to scale) if appropriate.

CONCLUSIONS:

1. What were the main distinguishing features which were used to differentiate between the classification types of :
    (a) Kingdoms?
    (b) within each Kingdom?

2. Comment on any difficulties found in this activity.

3. Why are classification systems often changed? Is the one used in the text the ultimately best system? If not, why not?

EXPERIMENT 12.5
Two lessons

EXAMINATION of SOME COMMON FOSSILS

AIM: To examine, describe and sketch some common examples of fossils.

MATERIALS: Sets of common fossils (one set/class group), hand lenses or stereo microscopes, pencils and coloured pencils, sheets of paper (photocopying paper is good).

BACKGROUND:

The most basic method of studying fossils is by a thorough examination and sketching of specimens. Considerable care must be put into the examination and sketch (photography is a poor alternative) so that all features which may give clues to the organism's type and environment can be seen. Usually, only a small fragment of the life-form is available and so palaeontologists often have to EXTRAPOLATE (extend data by educated guesses) their ideas using knowledge of modern organisms and their environment. They also use the modern view of life and its surroundings to make guesses about ancient times using the Principle of Uniformitarianism that states "the present is the key to the past".

PROCEDURE:

1. Carefully examine each fossil in turn looking for any internal features which can be identified.

2. Each specimen will have an identifying number or label to be written with the sketch and following description.

3. Draw the outline of each fossil in turn onto a clean sheet of paper to a satisfactory scale so that one sketch should take about one-half a page. Some specimens will have to be scaled up to this size by carefully copying the outline shape as near as possible to the original. Use x2, x3 etc. notation to indicate the scale.

4. Use coloured pencils to shade in the specimen by lightly shading using the side of the pencil then smearing it lightly with a finger. A 3D effect can also be given to the fossil by drawing a side around one edge and shading it with heavier pencil.

5. Review the internal structures of the specimen and sketch them within the outline.

6. Using the textbook, the Internet or other resources, identify each specimen in as much detail as possible and give the geological age (Period) to which these organisms belong.

7. Describe each fossil with a short paragraph indicating its shape, internal features and if possible the probable environment which would have supported such an organism (e.g. terrestrial, marine, freshwater etc.). They should also be classified as to their type e.g. plant, invertebrate (coral, echinoderm etc.) or vertebrate (reptile, bird, amphibian etc.)

EXPERIMENT 12.5   continued

DATA and OBSERVATIONS:

Detailed sketches, descriptions, classification and geological age for each specimen should be given in any order.

CONCLUSIONS:

Write a full conclusion with discussion about the observations made during this experiment.

1. Compare and contrast the use of originals and replicas in this study if appropriate. (Are replicas useful? Why?)

2. Why must the scale be given for each specimen?

3. Give uses for such a study of fossil specimens.

4. What are some problems with such a study? Could it be improved?

5. If this were a major work, what other research activities would be necessary to fully give more precise details of exact species, age and environment?

RESEARCH (Optional): Use the Internet or other resources to

1. Find out about how palaeontologists work in the (a) field and later in (b) the laboratory.

2. Where are the major fossil beds located nearby (with the State).

Chapter 13: Energy and the Earth

EXPERIMENT 13.1 One lesson

LAW of CONSERVATION of ENERGY

AIM: To demonstrate this law by converting mechanical energy into heat energy.

MATERIALS: One metre lengths of 50 mm internal diameter or less PVC tube, cork bungs or rubber stoppers or caps to seal the tube, 200 g lead shot, thermometer, beaker

BACKGROUND:

Mechanical energy consists of gravitational potential energy and kinetic energy. Gravitation potential energy (PE) is due to the object's mass (m), its height above a base line (h) and the acceleration due to gravity (g) i.e. PE = mgh. Kinetic energy is that of a moving object and it depends upon its mass and half of the square of its velocity i.e. KE = ½ mv^2. If an object with only potential energy is dropped from a height, just before it contacts with the baseline, all of this potential energy has been changed into kinetic energy. In this experiment, the lead shot (or steel ball-bearings) is dropped through a distance of 100 cm. 100 times. This continual change should raise the temperature of the mass of metal.

PROCEDURE:

1. Weigh out 100 g of lead shot and carefully pour it into the beaker with the thermometer. Measure its temperature.

2. Pour all of the shot into the PVC tube which has the other end firmly closed with a stopper.

3. Seal the open end with another stopper and holding both stoppered ends firmly, invert the tube violently 100 times (can be done as a relay between several students).

4. Quickly open one end and pour the lead shot carefully back into the beaker containing a thermometer and re-measure the temperature of the lead mass.

DATA and OBSERVATIONS:

1. Record the temperature of the lead shot before and after the experiment and therefore the change in temperature of the lead.

2. Use the specific heat of lead (128 J/kgK), the mass of lead (100g) and the measured difference in temperature to calculate the amount of heat change in the lead (note the units of specific heat is per <u>kilogram</u> mass).

3. Compare this to the theoretical value.

EXPERIMENT 13.1 continued

CALCULATIONS:

Energy lost by the mechanical energy of falling = heat gained by the lead

$$n(mgh) = mc\Delta T$$

where m = mass of lead (0.1 kg)
g = acceleration due to gravity (9.8 ms^{-2})
h = length of tube (1 metre)
n = number of inversions (100)
c = specific heat of lead (128 J/kgK)
ΔT = change in temperature (K equivalent to C°)

In <u>theory</u>, this experiment assumes that all of the mechanical energy will be lost to the lead shot to raise its temperature. By re-arranging the above equations:

$$n(\cancel{m}gh) = \cancel{m}c\Delta T$$

$$\Delta T = \frac{ngh}{c}$$

By substituting controlled variables: $\Delta T = \frac{100 \times 9.8 \times 1}{128}$

$$= 7.66 \text{ Celsius degrees.}$$

Of course this is a general experiment to show that mechanic energy can be changed into heat energy and the apparatus is fairly unsophisticated so there will be some errors.

CONCLUSIONS:

1. What was the change in temperature of the lead?

2. How did this agree with the theoretical value (of approx. 7.6 Celsius degrees)?

3. What was the error of this experiment expressed as a percentage error (i.e. absolute error/theoretical value x 100/1 %)?

4. Considering the errors, did the Law of Conservation of Energy operate? Discuss.

5. Account for this by listing some of the errors which may have occurred.

6. Are there ways of improving this experiment? Suggestions?

EXPERIMENT 13.2   Two or more lessons

ANAEROBIC and AEROBIC RESPIRATION

AIM: To investigate Aerobic and Anaerobic respiration.

MATERIALS: 250 ml conical flasks, 250 ml beaker, ceramic mat, cork and double-right angled delivery tube, test tubes, limewater, sugar, plain flour, dry yeast powder, warm water, thermometer, small evaporating basin or equivalent, drinking straws.

BACKGROUND:

Yeast is a single-celled organism which is classified as a type of fungus. It can encapsulate itself to retain moisture and is purchased in packets as a dry powder. Yeast obtains energy by anaerobic respiration of simple sugars producing as waste products carbon dioxide and alcohol. Sucrose is used in this experiment because it is convenient and easily obtained. Sucrose is a double sugar or disaccharide consisting of two bonded monosaccharides, glucose and fructose molecules (both $C_6H_{12}O_6$). In the first step of fermentation, an enzyme in the yeast breaks up the sucrose into a monosaccharide

$$C_{12}H_{22}O_{11} + H_2O \xrightarrow{ENZYME} 2\ C_6H_{12}O_6$$

Fermentation then breaks down the sugar through a series of complex reactions into carbon dioxide, alcohol (as ethanol) and energy. The overall reaction is:

$$C_6H_{12}O_6 \rightarrow 2\ C_2H_5OH + 2\ CO_2$$

The heat of this reaction ($\Delta H_{ferm.}$) is = -74.4 kJ i.e. an small exothermic reaction and it may be too small to measure if there is a great heat loss to the outside environment.

PROCEDURE: PART A:

1. Add about 150 ml of warm freshwater (about 35°C) to a 250 ml conical flask (use bottled water or distilled water if there is too much chlorine in tap water).

2. Add about 2 teaspoons (10 g) of sugar (sucrose) and about 1 g of dry yeast to the flask and shake it to mix the contents.

3. Measure and record the temperature of the mixture.

4. Place the cork with a double bend delivery tube into the top of the flask and insert the end of the delivery tube into limewater in a half-filled test tube ensuring that the end of the tube is well below the limewater surface as shown in the diagram. Limewater (calcium hydroxide solution) reacts with carbon dioxide gas to give a white suspension of calcium carbonate.

5. Half fill another test tube with limewater and place it upright next to the flask (it can be placed into the beaker at the side).

6. Place the apparatus and the extra test tube with limewater into a darkened, warm cupboard over night or until fermentation can be seen by bubbles of carbon dioxide in the flask.

EXPERIMENT 13.2 continued

PROCEDURE PART A : continued

7. Cautiously smell the fermenting liquid.

8. When the experiment has been completed, wash out all of the rest of the equipment.

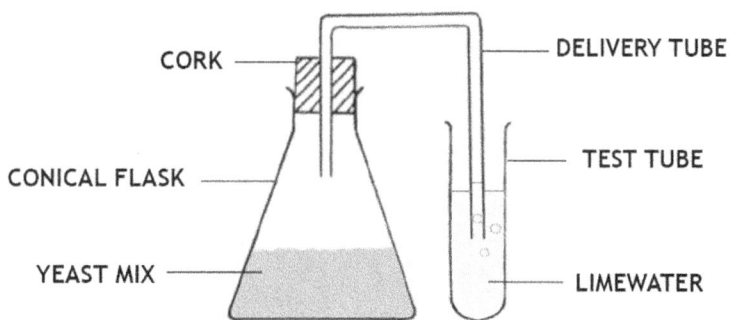

PROCEDURE PART B:

Half fill another test tube with limewater, insert the drinking straw and GENTLY blow into it. This breath is the result of aerobic respiration in the human body. Compare the result to another test-tube half-filled with limewater left to stand nearby.

DATA and OBSERVATIONS - PART A and B:

1. Record the temperature of the mixture before and after fermentation.

2. Describe the change in the limewater in the test tube with the delivery tube inserted and compare it to limewater in the extra test tube nearby.

3. Compare the limewater into which exhaled breath has been blown and compare it to the limewater in the test tube nearby.

PROCEEDURE PART C: (EXPERIMENT or DEMONSTRATION):

1. Take about 1 cup (120 g) of plain flour and thoroughly mix it with 2 teaspoons (4 g) of sugar and about half a teaspoon (2 g) of yeast powder in a small evaporating basin

2. Add water slowly and mix until a stiff dough can be formed into a small ball (i.e. it can be easily shaped in the hand and feels relatively dry).

3. Insert a small thermometer into the centre of the ball

EXPERIMENT 13.2 continued

PROCEEDURE PART C: continued

4. Make another dough ball as in the previous step but <u>leave out the yeast</u> (if done as a class experiment one group can make the ball with the yeast and the other group without the yeast).

5. Again insert a small thermometer into the centre of the ball.

6. Place both of these into a beaker, cover with clear wrap and place in a darkened cupboard.

7. After a time (say at least 15 minutes) remove the beakers and compare the temperatures.

DATA and OBSERVATIONS - PART C:

Compare the temperature inside the ball with the yeast to the ball without the yeast. Also note the room temperature.

CONCLUSIONS - ALL PARTS:

1. Was there any change in the temperature of the yeast mix in the conical flask in PART A? If not explain why. Compare this to any temperature changes PART C – was there any change to the temperature in the ball with the yeast? Explain.

2. Why was a flour ball made without yeast?

3. What did the change in limewater in both PART A and PART B show?

4. Why were separate test tubes of limewater placed nearby the experiments? Compare their colour with that of PART A and B.

5. Was there any change in the colour of the limewater which was set aside? If so, why did that occur? What was the purpose of this test tube?

6. Describe the smell of fermentation.

7. List all of the errors which could occur in this experiment.

8. Write a general statement about the two types of respiration observed.

OPTIONAL EXPERIMENT 13.3   One lesson

ENERGY AND THE WATER CYCLE

AIM: To show some of the heat transference within the Water Cycle using a laboratory model.

MATERIALS: Beaker, large watch glass, ice cubes, blue food dye (CARE: stains the skin), tripod, stand and Bunsen burner.

BACKGROUND:

In the Water Cycle, water goes through several changes in state or phase due to the transference of heat energy. This experiment uses a simple laboratory model to demonstrate some of these changes. CAUTION, the apparatus, water and steam will be very hot.

PROCEDURE:

1. Set up the apparatus as shown below and using water coloured with food dye to represent the sea. Half fill the beaker.

2. Place the large watch glass on top and start heating the water. Observe the interior of beaker and the base of the watch glass.

3. Before the water boils, place several (say 4-5) ice cubes into the watch glass.

4. Observe every detail of what happens.

5. Allow the apparatus to cool and then wash out the glassware thoroughly and put it away.

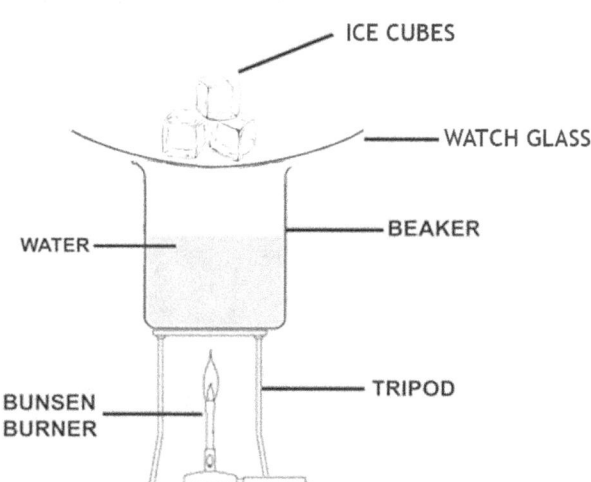

DATA and OBSERVATIONS:

Describe the results of the experiment noting <u>all</u> of the changes that occur, especially to the underside of the watch glass and with the ice cubes.

EXPERIMENT 13.3 continued

CONCLUSIONS:

1. What happens on the underside of the watch glass inside of the beaker? What colour is the substance which forms there? Explain.

2. What does this substance represent?

3. What happens to the ice cubes? Why? What do they represent in this model?

4. Comment on the use of this model to show some of the features of the Water Cycle.

Chapter 14: Energy and the Sea and Sky

OPTIONAL EXPERIMENT 14.1   One lesson

RADIATION and the INVERSE SQUARE LAW

AIM: To show the Inverse Square Law for energy radiation such as light and heat from the Sun.

MATERIALS: Lamp such as a microscope lamp or small infra-red lamp, photometer (or galvanometer/micro ammeter attached to a photocell), metre rules, stand for the photometer as required.

BACKGROUND:

Light, heat and other forms of electromagnetic radiation travel from the Sun to Earth with its intensity rapidly decreasing with distance such that Earth is in the Habitable Zone for living things. All radiation travels decreases rapidly with distance from its source such that the intensity of the radiation is proportional to the inverse of the square of the distance from the source. This is known as the Inverse Square Law i.e.

$$I \propto 1/d^2$$

Where I = intensity (several units used)
and d = distance

PROCEDURE:

1. In a darkened room, set up a lamp and place the photometer (or photocell attached to a galvanometer) at such a distance from the lamp that a maximum reading is obtained on the meter. Measure and record this distance with the meatre rule.

2. Move the photometer the same distance again i.e. doubling the distance from the lamp and again measure and record the reading on the photometer.

3. Again move the photometer out another length of the first distance from the lamp i.e. the photometer is now three times the distance out from the lamp than in step 1. Measure and record the value on the meter of the photometer.

4. Repeat this method with the photometer being place at four times the distance than that of step 1.

DATA and OBSERVATIONS:

1. Draw up a table for this data showing the distance from the lamp and the value of the units shown on the photometer or galvanometer.

2. Using an appropriate scale construct a graph showing this data using distance on the horizontal axis. Show all units and give a heading for this graph.

EXPERIMENT 14.1 continued

CONCLUSIONS:

1. What is the shape of the graph? What mathematical relationship does it show?

2. Does this graph show the Inverse Square Law for electromagnetic radiation?

3. What would be the errors in this experiment? How could they be removed or kept constant?

4. Explain, using a suitable diagram why the Inverse Square Law works.

EXPERIMENT 14.2	Two or three lessons

MODELING the GREENHOUSE EFFECT

AIM: To use a model to show the effect of greenhouse conditions on temperature.

MATERIALS: two or more rectangular 2 litre plastic bottles (e.g. cordial bottles), thermometers, large and flat glass or plastic dish which covers the surface of the bottles, black cardboard, white cardboard, rulers, bright electric lamp (at least 60 W globe), Alka-Seltzer antacid tablets, 250 ml beaker.

BACKGROUND:

The greenhouse effect is a natural process that warms the Earth's surface. When the Sun's light energy reaches the Earth's atmosphere, some is reflected back to space and the rest is strikes the surface of the Earth. Some of this is absorbed and some is reflected at lower energy radiation which cannot then go back into space. It is trapped within the atmosphere and absorbed by gases in the air including 'Greenhouse Gases' such as water vapour, carbon dioxide, ozone, nitrous oxide, methane and some artificial chemicals such as chlorofluorocarbons (CFCs). The absorbed energy warms the atmosphere and the surface of the Earth.

The three main effects which can cause temperature changes due to the Greenhouse Effect include the:

1. distance of the Sun from the Earth – as the planet travels in an elliptical orbit, sometimes about 94.5 million kilometres and sometimes 91.4 million kilometres;

2. albedo or reflectivity of the Earth which is about 30 – 35% on average depending upon cloud cover and the reflectivity of the surface. Dark surfaces have low albedo and white surfaces have high albedo; and

3. absorption by gases in the air, especially carbon dioxide and other Greenhouse gases.

This experiments attempts to use a model to determine if these factors change the temperature of the Earth's surface because of the so-called Greenhouse Effect.

PROCEDURE: Basic Control Model

1. Set up the control 'atmosphere' by drilling a small hole into the lid of a 2L plastic cordial bottle which has been washed, dried and its labels removed.

2. Insert a thermometer through this hole so that the reading of 20°C and upward can be seen on the outside. Seal the hole with plasticine or gum.

3. Measure the air temperature inside the bottle inside the classroom and record this temperature.

4. Go outside and find an open area exposed to direct sunlight and place the bottle flat upon a good, light coloured surface e.g. concrete (NOT bitumen).

EXPERIMENT 14.2 continued

PROCEDURE: continued.

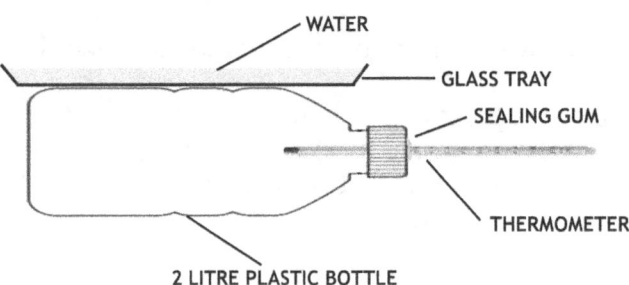

6. Leave this apparatus in sunlight for a set time (say 20 minutes) and record the change in temperature.

QUESTION: What is the purpose of the tray of water on top of the bottle? Discuss with the rest of the class before continuing with the rest of the experiment.

PART B: Distance from the Sun (DEMONSTRATION)

1. In a darkened room, set up the basic apparatus on a desk with a piece of black cardboard under the bottle.

2. Set up a strong lamp (at least 60 watts) above the dish (a sheet of glass on top of the dish might prevent the electric lamp touching the water) so that the bulb is about 20 cm from the surface of the water.

3. Record the temperature in the bottle then switch on the lamp and wait a set time (say about 20 minutes). Again record the temperature in the bottle. Record the data in a table of time and temperature.

4. Repeat this step with the lamp at 15 cm, 10 cm and on the glass. Record the temperature each time in the table.

5. Switch off the lamp and allow the bottle to cool.

6. Remove the lamp and tray and undo the cap from the bottle. Blow across the top of the bottle to force the hot air inside to flow out. Allow to cool and replace the cap. This step should be done to cool the air inside the bottle between all the following experiments.

EXPERIMENT 14.2 continued

PROCEDURE: continued.

PART B: Effect of Albedo

1. Again take the apparatus outside to a sunny location and lay the bottle on a <u>white</u> sheet of cardboard. Place the water tray on top and record the temperature immediately.

2. After the same time as before (say 20 minutes), measure and record the temperature.

3. Cool the inside of the bottle as outlined in Step 6 of Part A.

4. Restore the cap and thermometer and replace the white cardboard with a piece of <u>black</u> cardboard. Cover with the water tray as before and leave for the set time.

5. Record all of the temperatures in a table showing the colour of the cardboard and the initial temperature.

PART C: EFFECT of GREENHOUSE GASES - CARBON DIOXIDE

1. Fill the cooled bottle with about 300 ml of freshwater.

2. Ensure that the seal is good and lay the bottle on its side on top of black cardboard as before and place the water tray on top. Record the temperature now and at the end of the set time.

3. Cool the air inside as before but <u>leave the water inside</u>.

4. Now to make an 'atmosphere' of carbon dioxide by quickly adding at least 4 Alka-Seltzer tablets and then quickly screwing on the lid ensuring that the thermometer is well sealed. The tablets contain a solid acid and carbonate mixture which reacts with the water to produce carbon dioxide gas.

5. Place the bottle on its side again on top of black cardboard and replace the tray of water. Measure the temperature now and after the set time as before.

6. Record the temperatures in a table.

DATA and OBSERVATIONS:

These should consist of a sketch of the basic control apparatus, the temperature of the air inside the classroom and outside as well as tables for each of Parts A, B and C.

EXPERIMENT 14.2 continued

CONCLUSIONS:

Was there any change in temperature in the Control Experiment? What does indicate?

Was there any change in temperature in the bottle in Part A? What does this model suggest about the temperature effects due to the distance of Earth from the Sun?

Was there any difference in temperature when the 'surface' was changed from white to black? Explain.

Did the presence of additional carbon dioxide gas in the bottle change the temperature compared to the bottle with ordinary air? Explain.

This is a very basic model for what have been termed the Greenhouse Effect but there would be many problems with such a model. Suggest any problems with this model and if possible suggest some improvements (some research may be needed).

Could there be any other factors other than the three tested here that could change temperature through a Greenhouse Effect? Suggest how this model may be modified to test these hypotheses.

RESEARCH: Optional

Use the internet to look for other models which have been used to show the 'Greenhouse Effect'. Critically examine some of these models in a classroom discussion.

EXPERIMENT 14.3   One lesson

CONVECTION CURRENTS

AIM: To observe a model of a convection current in water.

MATERIALS: A large (1000 ml) Pyrex beaker, tripod and Bunsen burner (no gauze mat), heat mat for desk, crystals of potassium permanganate (Condy's Crystals), tweezers.

BACKGROUND:

Within the oceans, waters of different temperatures move around the surface and also below by convection currents and other factors.

PROCEDURE:

CARE should be taken with the Bunsen burner, hot tripod and any hot water in the beaker when cleaning up. Allow the tripod and beaker to cool where they stand before cleaning and returning equipment.

1) Fill the large beaker almost to the top with cold tap water and stand it carefully on the tripod.

2) Light a Bunsen burner, turn it to the blue flame reduced to a medium size.

3) Carefully take one single crystal (if they are small there may be several in about a cubic millimetre) of permanganate and carefully drop it down the side of the beaker and allow it to fall to the bottom.

4) Move the flame of the burner until it is directly below the crystal in the beaker.

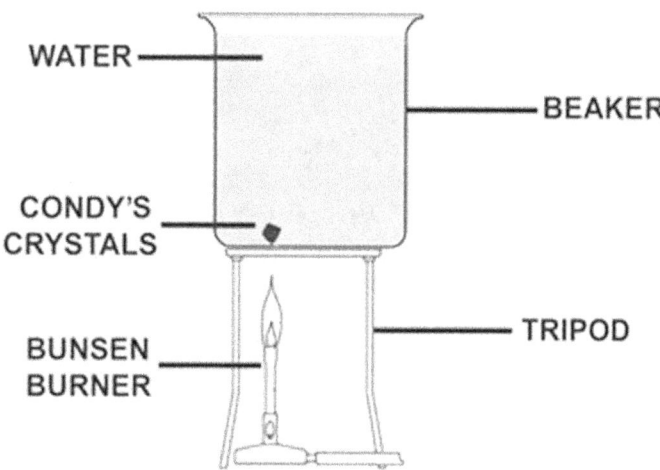

EXPERIMENT 14.3 continued

DATA and OBSERVATIONS:

Describe what has happened when the water was heated around the Condy's Crystals. Make a sketch of the resulting pattern after it has formed.

CONCLUSIONS:

1) What happens to the water:

    (a) immediately above the crystal

    (b) at the surface of the water and

    (c) on the opposite side of the beaker from the burner (cool side)?

2) What is a convection current?

3) How do convection currents operate in the world's oceans?

4) Given an example of one of the great convection currents in the ocean by naming:

    (a) a warm ocean current and

    (b) a cold ocean current.

EXPERIMENT 14.4 One or two lessons

PHOTOSYNTHESIS

AIM: To observe photosynthesis in an aquatic plant.

MATERIALS: Several *elodea* aquarium plants, beaker, large test tube, glass filter funnel, freshwater, sodium hydrogen carbonate (sodium bicarbonate) powder, coloured cellophane wrap (red, blue, green), accurate balance.

BACKGROUND:

Photosynthesis is the process by which green plants, including aquatic plants, manufacture simple sugars using water, carbon dioxide and sunlight. It is a complex process which occurs in the chloroplasts of plant cells involving the green pigment chlorophyll but an overall equation for it would be:

$$6CO_2 + 6H_2O + \text{light energy} \xrightarrow{\text{Chlorophyll}} C_6H_{12}O_6 + 6O_2.$$

PROCEDURE:

1. The following apparatus will be set up using a small sprig of elodea plant obtained from aquaria supplies. Make sure that the test tube is completely full of water before holding a thumb over the mouth and inverting it into the beaker of water over the funnel stem:

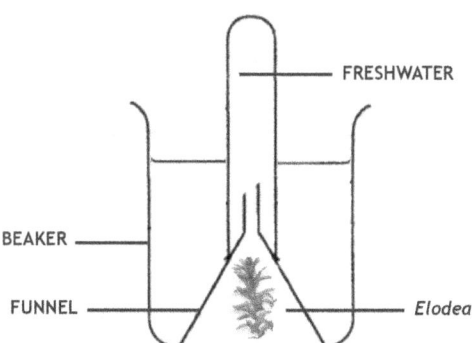

Several variations of this activity can be performed by different class groups using identical equipment and lengths of plant:

   a). Using this set up only with no additions just the water and its naturally-dissolved carbon dioxide.

   b). As above but increasing the carbon dioxide content of the water by adding about 0.5 g. of sodium hydrogen carbonate powder.

   c). As above but adding an extra amount of 0.05 sodium hydrogen carbonate (i.e. a total of 0.1 g).

EXPERIMENT 14.4 continued

PROCEDURE: continued

    d). The basic experiment (as in Step 1) but covering the whole apparatus in a layer of green cellophane.

    e). As above but using red cellophane.

    f). As for step b).but using blue cellophane.

2. If possible, all experiments should be placed in direct sunlight. If not possible then several bright fluorescent lamps can be used.

3. Each set up is observed carefully and the number of bubbles coming off the plant and up into the test tube are counted over time – say every five minutes for about 20 minutes.

4. Number of bubbles and the time periods should be recorded in a table.

5. If there is sufficient gas in the test tube (say about one quarter or more) then the test tube can be carefully removed by placing a thumb over its mouth whilst still below water level, taken out, turned upright and then the gas tested by plunging a glowing extinguished match or previously lit toothpick quickly into the tube.

DATA and OBSERVATIONS:

2. Record the number of bubbles for the times for the experiment conducted.

3. Look at the group results and record the finding for each experiment in another table giving the various conditions over the maximum time (say 20 minutes) and the number of bubbles i.e.

	Basic	0.05 g 0.1 g sodium hydrogen carbonate		Green Red Blue cellophane filters		
Number of Bubbles						

4. Make a sketch of the basic apparatus i.e. Step 1a. after the total time with appropriate labels.

5. Describe any results of the test of the gas inside the test tube.

EXPERIMENT 14.4 continued

CONCLUSIONS:

5) Was there any difference between the number of bubbles over the total time for each of the different variations? If so:

   (a) which one gave the greater number of bubbles (gas volume)? Explain.

   (b) which one gave the least amount of bubbles? Explain.

6) If the gas was tested, what gas was it? Why?

7) What other factors could be changed to determine how photosynthesis works?

8) What errors could have occurred in the experiment?

9) Write a general statement about photosynthesis in plants and its requirements.

EXPERIMENT 14.4 continued

EXPERIMENT 14.5   One lesson

POPULATION GROWTH of a BACTERIA COLONY

AIM: To observe the rate of population growth in bacteria.

MATERIALS: Time lapse video of bacteria growth in a petri dish (from YouTube by Serafina C), graph paper.

BACKGROUND:

Population studies are often difficult because most species take a long time to reach their climax population and habits are often large or widely spread. Bacteria cultivated in a small Petrie dish are ideal because they reproduce (by asexual fission) rapidly with cell division about every 20 minutes. Moreover, they can be counted relatively easily under a microscope over time.

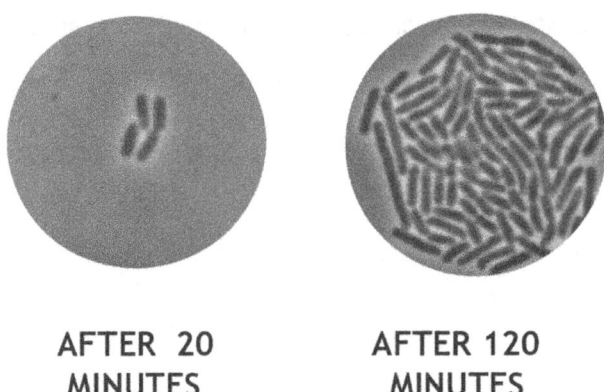

AFTER 20 MINUTES   AFTER 120 MINUTES

This exercise is simple and does not have the added dangers of dealing with complex procedures of setting up or the dangers of dealing with potentially harmful bacteria. Unfortunately the conditions are laboratory conditions and the video does not show the colony reaching its Carrying Capacity so only the first part of the Logistics Curve can be drawn.

PROCEDURE:

Beforehand, construct a table showing number of bacteria over several time periods. Watch the video either individually or as a group, stopping the action as indicated on the video.

https://www.youtube.com/watch?v=KIpcCyuypzg

1. At each pause, count the number of bacteria seen in the field of view and record the time and number in a table.

2. At the end of the video, construct a graph of bacteria population (horizontal axis) against time (vertical) axis.

3. Use the best slope of the graph to determine the rate of growth of the colony (number/time).

EXPERIMENT 14.5 continued

DATA and OBSERVATIONS:

1. In the table, record the number of bacteria counted at each time slot.

2. Draw a graph of this growth rate

CONCLUSIONS:

1. Describe the shape of the final graph.

2. Mathematically, what relationship does this graph show?

3. Is it consistent with the usual Logistics Curve for populations in Nature?

4. How could this (original) experiment be expanded so that the bacteria can reach their climax population?

5. In the petrie dish, what factors would limit the growth of the bacteria?

APPENDIX A: Risk Assessment of Practical Work and Excursions

It has become mandatory that a risk assessment is made before carrying out any laboratory practical activity or excursion. These must consider:

1. the nature of the activity;

2. equipment, chemicals, living things and other items to be used;

3. the nature of the environment, especially if venturing outside for an excursion, even in the local area; and

4. the student.

Once all of the factors and options have been considered, a Risk Assessment Form should be completed and kept in a central records section (often kept by a Laboratory Manager). It is also a good idea for a copy to be kept (electronically and print) by the teacher along with the set of notes for the experiments and activities (perhaps attached to the teacher's cop[y of this book). Once completed, it should be revised and modified as required each time the activity is performed. This need not be an onerous task after it has been done for the first time. Thereafter it simply requires a quick review of the previous form and modifications if any of the variables (e.g. different student type, different location, new equipment, different teacher etc.) have changed.

Safety is paramount for any teaching activity and the students must be made aware of any potential hazard in the activity and take adequate precautions. Some suggestions for laboratory safety and safety during field excursions have been given at the beginning of this book. It is always useful to have these suggestions displayed prominently in the laboratory. In the associated textbook ADVENTURES IN EARTH and ENVIRONMENTAL SCIENCE, each chapter contains a PRACTICAL TIPS at its conclusion. These have been based on over forty years of safe laboratory and field practices as a field researcher well as many years travelling into some of the most hostile and remote parts of the worlds, such as the Antarctic Peninsula, Amazon Basin, north African deserts and alpine regions in New Zealand, Europe and the Andes. This section also outlines, as does various, parts of the body of the textbook, how science is one of the most interesting endeavor of Humankind.

Institution laboratory managers should have on file a detailed indexed catalog of the risk or hazard potential of all chemicals, electrical equipment, biological specimens (including live organisms) and other items in use within the laboratories.

A simple Risk Assessment Form is given on the next page:

Earth and Environmental Science
Risk Assessment Form

DATE:	CLASS:	LOCATION:	
ACTIVITY:		TYPE (Pract./Excursion/visit):	
RISK LEVEL: (High/Medium/Low)	Students:	Teacher:	Other(who/what):

OUTLINE of ACTIONS	RISK	PRECAUTION/ACTIONS

DISPOSAL of WASTES:

OTHER COMMENTS:

A more detailed Risk Assessment Form can be found at:

https://www.aisnsw.edu.au/workplace-health-and-safety/Documents/Appendix A Science and Technology Risk Assessment Template Rev.1.docx

Some useful references for risk assessment can be found at:

https://education.qld.gov.au/sitesearch/Pages/results.aspx#k=risk%20assessment

https://education.qld.gov.au/initiatives-and-strategies/health-and-wellbeing/workplaces/safety/managing/risk-management

https://www.riskassess.com.au/info/routine_safety_procedures

https://smah.uow.edu.au/content/groups/public/@web/@sci/@chem/documents/doc/uow016874.pdf

https://www.riskassess.com.au/docs/RABrochureAU.pdf

http://www.nswtitration.com/files/school_risk_assess.pdf

https://assist.asta.edu.au/

https://assist.asta.edu.au/search?expert=&field_curriculum_year=All&field_publication_date=All&keywords=risk+assessment&field_tax_australian_curriculum_parent_parent_parent_tid=&laboratory_technicians=All&area=&field_voting_user_rating=&field_voting_average_rating_1=&field_voting_user_rating_1=&rating_point=All&year_level=&sort_by=created

https://education.nsw.gov.au/teaching-and-learning/curriculum/key-learning-areas/science/safety

https://www.education.vic.gov.au/school/principals/spag/governance/Pages/riskinplanning.aspx

http://ascip.org/wp-content/uploads/2016/06/ASCIP-Risk-Management-Primer-for-School-Districts-SIXTH-DRAFT-2016-06-20.pdf

https://www.rospa.com/rospaweb/docs/advice-services/school-college-safety/managing-safety-schools-colleges.pdf

https://www.leeds.ac.uk/secretariat/documents/risk_management_guidance.pdf

https://www.teachers.org.uk/files/safety-in-practical-lessons_0.doc

APPENDIX B: Excursion Permission Note

Some institutions often require an Excursion Permission Note from students who are under legal age who are leaving the institution for a field, industrial/scientific visit or other outside activity. These DO NOT absolve the teacher and guides of any moral and legal responsibility but merely provides parents/guardians with the necessary information as to where and what their child will be doing. It also can provide a list of safety and comfort items which will be needed from home for the activity in the hope that such reasonable items will be supplied.

Some examples of field or other outside activities for this program include:

- Rock Quarries
- Mining areas (special permission and rules from the company)
- Freshwater stream ecology
- Marine Rock Platform ecology
- Rainforest/Dry Forest/Grasslands ecology
- Mining/mineral Museums
- Environmental Stations (special rules and guides apply)
- Museums of Natural History
- Scientific Research Organisations (special rules and guides apply)
- National Parks (special local rules and guides necessary)
- Government Agencies (special rules and guides apply)

A typical activity Permission Note is given on the next page and may be copies for the students to take home:

< insert institution letter head>

EARTH and ENVIRONMENTAL SCIENCE EXCURSION

DATE: <insert date and times> **CLASS:** < insert class/group>
DESTINATION: <insert location/name of place or organisation etc.>

This excursion is an integral part of the semester's Programme. All students will be required to complete an assignment, associated with the excursion, which will contribute towards the assessment of the subject.

ARRANGEMENTS: < insert details of transportation, time of departure and return etc.>

The bus will leave at   < time >   and returns to the school at   < time >   approx.
(Bus company can be contacted at: <name and contact number>   )

Students are to meet at**:**   < meeting place and time>
but will not to enter the bus until directed to do so. The rules of the School apply at **all times**

Staff going on the excursion will be: < name(s) of staff and contact numbers >

SPECIAL REQUIREMENTS: < special requirements such as dress, personal items such as cameras, mobile phones, writing material, safety precautions etc. Also any special rules of behaviour and group actions if lost etc.>

Yours faithfully,
 < name of person in authority>

Head of Earth and Environmental Science

CONSENT FORM: Please return to supervising teacher by   < insert time/date>

I have read the above information and agree for...... **<insert student name>** ....to go on the excursion.

Special information concerning my daughter's/son's welfare that the supervising staff should know is as follows:

<insert special requirements such as allergies, dietary requirements, medical conditions etc.>

Phone contact of parent/guardian   Home:..........................Work.............................

Signature:..(Parent/Guardian)

Date ……………………………………….

Books by the Author

ADVENTURES IN EARTH AND ENVIRONMENTAL SCIENCE - BOOK 1. This is the first book of two which looks at the Earth, its matter, energy relationships and its life and how all interact together as a whole. The Earth is seen as a closed system contain the Earth's materials and living things but allowing a necessary flow of energy into and out of the planet. The atmosphere, hydrosphere, geosphere and biosphere of Earth are all examined in detail and lavishly illustrated with over five hundred photographs and diagrams. There are also several links to videos made by the author during his own adventures in studying the Earth. Each chapter is concluded with a Summary, Practical Tips, ten Multichoice Questions and ten longer Review and Discussion Questions. There is also an accompanying PRACTICAL MANUAL with a large number of experiments and data analysis activities which can be performed by students using basic available equipment in support of the textbook. This manual also teaches students how to investigate and write research reports for submission.

ADVENTURES IN EARTH AND ENVIRONMENTAL SCIENCE - BOOK 2. This is the second book with this title and, after a short revision chapter, looks at how Humankind lives on planet Earth. Renewable and non-renewable resources are described and how their use has impacted on the world's ecosystems, on land, in the sea and in the atmosphere. This is also discussed with an emphasis on the problems of future energy needs, global warming and social consequences of these events. As well as problems caused by Humankind, the natural hazards of the Earth have also been described with the view that many or the world's populations live in regions that can be very dangerous at times. The contents of this book are also supported with many photos, illustrations and videos. Each chapter is concluded with a Summary, Practical Tips, ten Multichoice Questions and ten longer Review and Discussion Questions. Book 2 also has an accompanying PRACTICAL MANUAL with a large number of experiments and data analysis activities which can be performed by students using basic available equipment in support of the textbook.

ADVENTURES IN EARTH AND ENVIRONMENTAL SCIENCE is the composite book containing all of the content of Books 1 and Books 2. It has been written as a utilitarian reference book for the classroom, library or home study.

ADVENTURES IN EARTH AND ENVIRONMENTAL SCIENCE - TEACHERS' GUIDE has been designed to assist teachers with the use of these books and the teaching of this subject. It gives advice on lesson preparation, the teaching of the practical work and answers to the questions contained in the books.

Adventures in Earth and Environmental Sciences Books 1, 2 , the composite book and the Teachers Guide are all available in electronic (Kindle) format which can be viewed using any electronic device having the free Kindle App. They are also available in PRINT editions from Felix Publishing at:

(info@felixpublishing.com)

Other books in the **ADVENTURES** series follow a more traditional Earth Science content in the sciences of Geology, Oceanography, Meteorology and Astronomy. They are also available in electronic format from Kindle as well as in print form direct from Felix Publishing (info@felixpublishing.com) and include:

ADVENTURES IN EARTH SCIENCE ADVENTURES in EARTH SCIENCE is an in-depth textbook as well as a series of adventures across seven continents and beyond in the sciences of astronomy, geology, meteorology and oceanography. It has been written with over forty years of experience in studying, researching and teaching earth science. Whilst it has been designed for senior high school and junior university or college, it is written in an easy style and well-illustrated so that anyone with an interest in this topic would find it an interesting and valuable resource.

This reference book has also been reprinted in the smaller A5 size as separate topic editions for easier reading by anyone wishing to have the most up to date information on these topics. The smaller books are:

EXPLORATION SCIENCE
Field Geology & Mapping

RICHES from the EARTH
Minerals & Energy

CHANGING the SURFACE
Erosion & Landscapes

ROCKS - BUILDING the EARTH

FOSSILS - LIFE in the ROCKS. Fossils and past environments

A DANGEROUS PLANET
Volcanoes & Earthquakes

THROUGH SEA and SKY
Oceanography & Meteorology

BEYOND PLANET EARTH
An Introduction to Astronomy

For more information about all of these books and world-wide distribution contact the publisher direct at info@felixpublishing.com

www.ingramcontent.com/pod-product-compliance
Lightning Source LLC
Chambersburg PA
CBHW050714090526
44587CB00019B/3367